On the Waters of the Wissahickon

On the Waters

A History of Erdenheim Farm

Eric Plaag

of the Wissahickon

THE UNIVERSITY OF SOUTH CAROLINA PRESS

Published in Cooperation with the University of South Carolina College of Arts and Sciences

Published by the University of South Carolina Press
Columbia, South Carolina 29208

www.sc.edu/uscpress

Manufactured in China

24 23 22 21 20 19 18 17 16 15 10 9 8 7 6 5 4 3 2 1

Library of Congress Cataloging-in-Publication Data can be found at http://catalog.loc.gov/.

ISBN: 978-1-61117-549-3 (cloth)
ISBN: 978-1-61117-550-9 (ebook)

FRONTISPIECE: Erdenheim's Stone Bridge over the Wissahickon.
Photo by Eric Plaag, May 2012.

CONTENTS

INTRODUCTION

It sits on the outskirts of Philadelphia, straddling the boundary between Whitemarsh and Springfield Townships, in the valley of the Wissahickon, a meandering ribbon of water known by the Delaware Indian tribe that occupied this land in 1677 as the "Wicssahitkonk" or "catfish or yellow-water stream."[1] Less than a mile to the east are the plains on which the British and the American colonists fought the Battle of Whitemarsh, a crucial conflict in the struggle for American independence. To the south are the old towns of Germantown and Chestnut Hill, places where the earliest German immigrants first found a foothold on American shores, seeking religious freedom and liberty. All around, in every direction—and indeed on its very soil—are the vestigial remnants of the Quaker ideals and ways of life that shaped what was once the colonial backcountry, then—after 1784—a rich and productive collection of farms and hamlets on the eastern end of Montgomery County.

For at least the past 150 years—and perhaps another century longer—it has been known as Erdenheim, a name that many assume to be German for "earthly home" or "home on earth," and the fact that the place was owned 250 years ago by a German immigrant, Johann Georg Hocker, who made a deeply satisfying life for himself in America, only adds to the perceived authenticity of its name and the enduring charm of the mythology that surrounds the place. Over the long arc of its history, many of the great legends of horseracing—Flora Temple, Leamington, Iroquois, Parole, Jaipur—trained in its paddocks and foaled countless champions after them in Erdenheim's stalls. Its owners have been from the oldest families in Whitemarsh, Springfield, and Montgomery County—Farmar, Robeson, Streeper, Williams, Hocker, Coulston, Scheetz, Corson, and Lukens. Today the mere mention of Erdenheim brings a wistful look to the eyes of local residents and even those who commute from afar through Whitemarsh to jobs in Philadelphia; many folks claim to drive several miles out of their way, far from the interstate and the bigger highways, just for the chance to ride alongside Erdenheim's fields and maybe catch a glimpse of a prized stallion, bull, or flock of sheep, all of which have made Erdenheim famous, both locally and internationally, for centuries.

Erdenheim's modern boundaries are fairly straightforward. To the south are Northwestern Avenue and Wissahickon Avenue, a single road that crosses Stenton Avenue and marks the boundary between Philadelphia and Springfield Township's panhandle along the Northwestern Avenue stretch (above which are the Whitemarsh Valley Country Club and the Bloomfield Farm portion of Morris Arboretum, then Erdenheim north of that), with Erdenheim's Equestrian Tract and Carson College north of the Wissahickon Avenue segment. Carson College accounts for the majority of Erdenheim's eastern boundary. Stenton Avenue runs north from Wissahickon and Northwestern Avenues, dividing Erdenheim's Equestrian Tract from its Wissahickon Tract, although historically these lands were part of the same parcel. To the north of the Equestrian Tract is West Mill Road, which leads here from nearby Flourtown and meets Stenton Avenue at a curve, where Stenton Avenue continues on to create the rest of

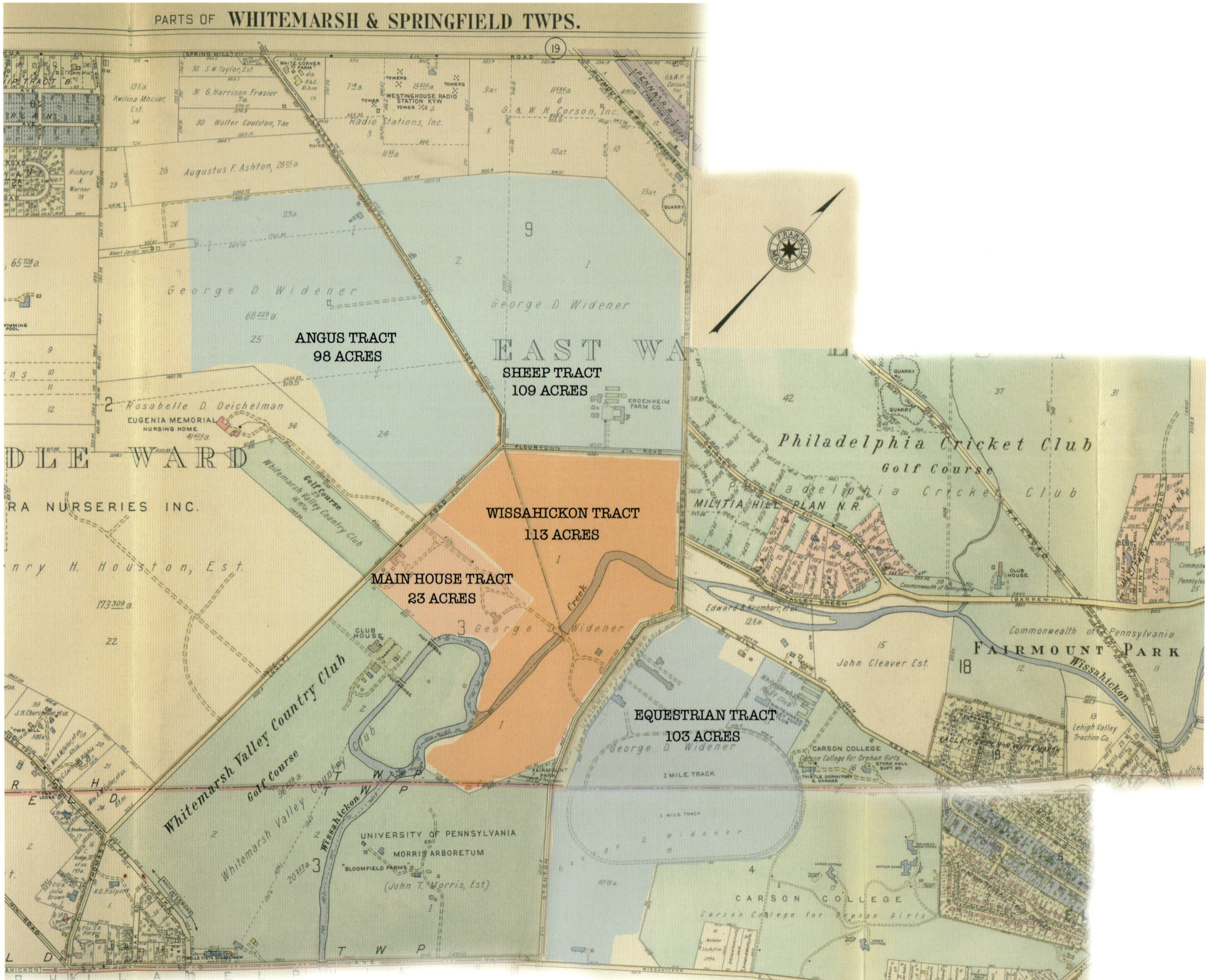

PARTS OF WHITEMARSH & SPRINGFIELD TWPS.
ANGUS TRACT
98 ACRES
SHEEP TRACT
109 ACRES
WISSAHICKON TRACT
113 ACRES
MAIN HOUSE TRACT
23 ACRES
EQUESTRIAN TRACT
103 ACRES
George D. Widener
EAST WA
Rosabelle D. Deichelman
EUGENIA MEMORIAL NURSING HOME
Whitemarsh Valley Country Club
Golf Course
Philadelphia Cricket Club
Golf Course
MILITIA HILL PLAN N.R.
FAIRMOUNT PARK
Commonwealth of Pennsylvania
John Cleaver Est.
CARSON COLLEGE
Carson College for Orphan Girls
UNIVERSITY OF PENNSYLVANIA
MORRIS ARBORETUM
(John T. Morris, Est)
Wissahickon
WESTINGHOUSE RADIO STATION KYW
Radio Stations, Inc.
Augustus F. Ashton
G. & W. H. Corson, Inc.
ERDENHEIM FARM CO.
FLOURTOWN ROAD
RA NURSERIES INC.
nry H. Houston, Est.
DLE WARD
1 MILE TRACK
Edward B. Krumbhaar, et al
Lehigh Valley Traction Co.

Erdenheim's northern boundary. To the west is Thomas Road, running north from Northwestern Avenue then crossing into Erdenheim's territory, where it terminates at Erdenheim's geographical center—the intersection with Flourtown Road. This latter byway extends to the east and west, dividing Erdenheim's Wissahickon, Sheep, and Angus Tracts from one another and leaving Erdenheim to the west just below Joshua Road. Meanwhile the Springfield-Whitemarsh Township boundary runs through the southeastern extremity of modern Erdenheim Farm, bisecting its historic one-mile oval.

It would be a mistake to think of Erdenheim as historically being one farm, though. It isn't now, technically speaking, and it never really was. The farmstead that many locals called Erdenheim for the better part of two centuries was actually a much smaller farm, for many years just over 155 acres, but at times perhaps as small as thirty-two, nestled on both sides of present-day Stenton Avenue and almost entirely on the east side of the Wissahickon. Before 1916 the parcels today known as the "Sheep Tract" and the "Angus Tract" were divided into numerous lots and owned by multiple owners, none of them the owner of the place historically called Erdenheim, or what will be called "Old" Erdenheim throughout most of this history. The parcel known these days as the "Main House Tract"—what most locals today believe is the historic plantation tract for Erdenheim—wasn't even a part of Erdenheim's holdings until the late nineteenth century, which explains why this history most frequently refers to it as "New" Erdenheim. The parcel now known as the "Wissahickon Tract" is equally misleading in name and boundary, given that the land on the west side of the Wissahickon was never a part of "Old" Erdenheim, while the portion of this tract on the east side of the Wissahickon always was. And present-day Erdenheim's final tract—the "Trackside Equestrian Tract"—is also deceptive as evidence of Erdenheim's past, given that its boundaries encompass portions of "Old" Erdenheim, including parcels that Hocker at one time sold off and other owners brought back into Erdenheim's holdings at a much later date, but exclude vast swaths of land that were at one time part of "Old" Erdenheim and now belong to others.

In many respects these modern parcel divisions are a consequence of the division of land that occurred following the death of its principal owner during the twentieth century, George D. Widener Jr. After Widener passed in 1971, 117 acres of Erdenheim's then nearly 450 acres were bequeathed to the Natural Lands Trust for conservation. The remainder went to Fitz Eugene Dixon Jr., Widener's nephew. Upon Dixon's death in 2006, many feared that the Dixon holdings would be divided up into lots for subdivisions and other development, as most of the farmland near Erdenheim had been handled during the twentieth century. Through the efforts of the Whitemarsh Foundation, Peter and Bonnie McCausland, and numerous other agencies and contributors, the five tracts that today make up Erdenheim were conserved through a series of elaborate cooperative agreements among several parties that culminated in the present arrangement by June 2009—a process that will be described in greater detail toward the end of this history.

Discovering Erdenheim

I first heard about Erdenheim Farm in December 2011, when Mary Anne Fitzpatrick, dean of the College of Arts and Sciences at the University of South Carolina (USC), and Dr. Lawrence Glickman, chair of the History Department at USC, contacted me about preparing a history of the farm for them. The timeline was short; Dean Fitzpatrick and Dr. Glickman asked that I find out everything I could about Erdenheim over a six-week period, including a one-week trip to the outskirts of Philadelphia, then report back immediately on the feasibility

FACING: Composite map from plates 3, 17, and 20 of *Property Atlas of Montgomery County*, volume D, 1949, from Map Collection, Free Library of Philadelphia, with overlay of current Erdenheim Farm tracts. Overlay boundaries are approximate and for illustrative purposes only.

of a book-length project on the farm. Given that I knew almost nothing about horseracing, sheep, or cattle, next to nothing about the townships on the outskirts of Philadelphia, and nary an iota about Erdenheim, I was, to say the least, a bit daunted by the task. Then they threw in the kicker—under no circumstances was I to contact the farm, the Whitemarsh Foundation, or Peter and Bonnie McCausland, Erdenheim's current owners. This proposed book was intended, after all, to be a thank you of sorts. A gift. A surprise.

That first week in Montgomery County in January 2012 was something of a whirlwind. Armed with some (very) basic "facts" about the history of Erdenheim, as gleaned from the farm's website and a handful of other online resources, I showed up at the Historical Society of Montgomery County, hoping to tease out some more background. There were a few clippings about the farm and some of its more famous owners—George D. Widener Jr., Aristides Welch, Robert N. Carson—scattered in their collections, but for the most part, I still felt adrift in a sea of ignorance about the place. I took a drive out to Erdenheim, photographed several of its properties from the road, and tried to get a sense of the land and its spirit. I spent the next afternoon wandering from one local historical society to another—some of them open and very helpful, others all but defunct—hoping to elicit from their archivists all the data they could muster about Erdenheim and its seemingly enigmatic past.

That's when the stories started. The myths. The legends. I heard a rumor that George Washington had once slept there, and maybe the Marquis de Lafayette, too. I heard that the farm had been at the center of the lines in the Battle of Whitemarsh during the American Revolution, and that its owners had given quarter and supplies to American troops. I heard that the property had been an inn once, under the proprietorship of the man alleged to be Erdenheim's first owner, George Hocker. Then I heard about the other presidents who had allegedly spent the night there during the nineteenth century—first it was just James Buchanan, then Buchanan and Grant, and eventually a total of eight U.S. presidents, although no one could quite remember who they all were. Some of the stories got even wilder. There were tales of suicides at Erdenheim, lots of them, and winking hints that insurance arson was behind some of Erdenheim's many fires. One person told me that George Hocker had once killed an intruder on the premises. Another librarian told me that there were bodies secretly buried on the hilltop overlooking the sheep farm and the Wissahickon. Still another said that one of the owners had died on the *Titanic* and left half the farm and most of his money to form an orphanage for young girls. I even heard passing references that one of the Erdenheim owners—allegedly a well-known fraud and con man—had shot an actress in cold blood on a Philadelphia stage, infuriated that she had left him for another man. The real doozy came when three separate individuals at different archives told me that Sirhan Sirhan, the infamous assassin of Robert Kennedy, had worked as a stable hand and jockey at Erdenheim just weeks before he killed Bobby.

I will tell you now that there is a core of truth in each of these stories, and in the history that follows, you'll hear a bit about each of these tales, along with the actual facts that help us understand the truth of the matter and how Erdenheim fits into it. There will also be some revelations that might be surprising and a little difficult to accept; local tradition, for example, holds that Erdenheim's original manse, dating to at least 1764, is still sitting out there in a field standing guard over the glorious Wissahickon, but the truth of the matter is that "Old" Erdenheim is long gone, demolished nearly a century ago for reasons that

FACING: Twentieth-century horse stable and exercise barn on east side of Stenton Avenue, where Sirhan Sirhan is alleged to have worked, January 2012. Photo by Eric Plaag.

remain lost to history. This is the lesson: Tradition and fact, sadly, do not always align. Instead local historical traditions are sometimes like a game of Operator. One person digs up a nifty fact and shares it with another. That person either mishears that fact, or misremembers it, or simply manufactures some additional details to make it more juicy, then passes it along to another equally flawed storyteller. Then folks invent additional layers of history in order to better hold on to a past that has vanished. Eventually what remains no longer resembles the story that began the cycle. To illustrate how this process works, it might be worth discussing just one of those tales about Erdenheim here, right now, so that we can forever dispense with its absurdity but also understand its origins.

Sirhan Sirhan and the Stables at Erdenheim

Because it was one of the more persistent and fascinating stories I heard the week of that first visit to Erdenheim in January 2012, I really wanted to pin down the facts, thinking it might serve as an excellent starting point for this book. As it turns out, though, most folks who have written about Sirhan Sirhan aren't very interested in his life prior to the assassination, except to the extent of digging up anti-Israeli sentiments he might have harbored as the genesis of his hatred for Robert Kennedy. There are many sources out there that discuss the events of that horrible day, or the meaning and consequences of the assassination, or even the geopolitical implications of the event, but almost no one seems to talk much about the minutiae of the life of the man accused of pulling the trigger. Part of this may be because Sirhan himself apparently spent the first few years after his arrest claiming to be a very different kind of man than he was prior to the assassination. Thankfully, though, a reporter named Aziz Shihab, who penned a brief, not particularly well-written, and ridiculously difficult to find biography of Sirhan just months after the assassination, was kind enough to include a short chapter on Sirhan's life prior to 1968.[2]

As with each of the other wild tales about Erdenheim referenced earlier in this introduction, there is something of a core of truth to the rumors about Sirhan's connection to Erdenheim. He was, indeed, a man fascinated with horses in the months leading up to Kennedy's assassination. After immigrating to the United States as a young boy with his family in 1957, and spending almost the entirety of his American youth in Pasadena, California, Sirhan graduated from John Muir High School in 1963. He took two years of courses at Pasadena City College. And in the afternoons following those classes, he parked himself in the stands at the Santa Anita racecourse, betting every dollar he had on the ponies. Sometimes he went home with what, to him, must have seemed a small fortune. Other days bad bets cleaned him out.

Like many young men in the 1960s, Sirhan was not a guy who seemed very clear about his purpose in life. In September 1964 he took a job at a service station on the outskirts of Pasadena, and when he got tired of that job, he found another one at the service station on the opposite corner. He kept to himself, mostly, and when he wasn't at work, he spent most of his free time pondering his next move. He still went to play the ponies on a regular basis. And one afternoon, when he came home with a wad of cash like he'd never seen before, he quit that second gas station job, too. When he got bored a few weeks later, he turned to an elderly gardener who worked close to the service stations, filling in as a handyman and plant waterer. That job didn't last long, either.

The money from the tracks seemed to change things for him, though. Now, it seemed, there was purpose in his life. He registered with the California Horse Racing Board in 1965 and 1966 as a training jockey and an exercise boy, filling in where needed at the courses at Santa Anita and Hollywood Park. During the summer he found work at the Granja Vista Del Rio Horse Ranch in Corona, continuing his promising run as an exercise jockey. Then, just as quickly as they had changed for the positive at the track two years earlier, his fortunes shifted again at the ranch. He was thrown on September 24, 1966, suffered what must have been some sort of head trauma (his vision was badly affected), and filed a workers' compensation case, for which he won a $2,000 settlement. After being out of work for nearly a year, he found a new job away from the horses, this time working for an organic grocery store near his home. Disappointment turned to anger. The girl he claimed to love found a new lover and rarely spoke with him. His new boss argued with him frequently over events in the Middle East. On March 7, 1968, after weeks of quibbling with the shop owner about Jews and Arabs, Sirhan had enough. He quit his job, spent the next three months in Pasadena pining over that old girlfriend, then shot a presidential candidate to death on June 5, 1968, in what author Shihab says was a blind, jealous rage over a broken heart.

The name of Sirhan's employer, though, is what stands out for our purposes: John Henry Weidner.[3]

To the casual observer on the outskirts of Philadelphia, the genesis of the Sirhan-Erdenheim rumor is easy to trace. In the game of historical Operator that so often gets played in our rumor-mongering, scandalized, and scandalizing culture, "Sirhan was an exercise jockey, and he worked for a man named Weidner just before he killed Bobby Kennedy" easily becomes "Sirhan was an exercise jockey for George Widener at Erdenheim before he went to California and killed Bobby Kennedy." But John Henry Weidner was not George D. Widener Jr., and there is not one single shred of evidence to suggest that the two men ever knew one another or were related to each other, nor any evidence that Sirhan spent any time outside of California in the years leading up to the assassination. He most certainly was never at Erdenheim, and in spite of his brief but committed flirtation with the horseracing world, he may not have ever heard of the place.

In the circuit of second- and thirdhand knowledge, however, those aren't the details that are memorable, and those kinds of wild stories tend to stick after a while, the facts be damned. And to a local from Whitemarsh, it's much more interesting to play the old saw's tune that Sirhan once worked at that farm down the street. As I learned firsthand, it makes for good conversation.

I'm not a local from Whitemarsh, though. I'm a professional historian. My job is to set the record straight.

Putting the Real Story Together

Histories like this almost always have a single author's name on the title page, but it would be a terrible mistake to suggest that everything here is strictly my work. Beyond the various primary and secondary sources created by newspaper reporters, historians, archivists, and the historical figures themselves, this history reflects the hidden hands of dozens of individuals who gave selflessly of their time, energy, and intellect to help me accumulate and make sense of the bins of paper and gigabytes of information that tell Erdenheim's story. It seems only fair to call them out here and ask that they take a bow.

During the research phase of this project, I visited dozens of repositories all over the country. In particular I am grateful to the librarians, archivists, and general staff of the American Antiquarian Society, the Belk Library at Appalachian State University, Chestnut Hill College, the Chestnut Hill Historical Society, the Frederick Law Olmsted National Historic Site, the Free Library of Philadelphia, the Historical Society of Fort Washington, the Historical Society of Montgomery County, the Historical Society of Pennsylvania, Keeneland Library, the Library Company of Philadelphia, the Library of Congress, the Montgomery County

Archives, the New York Historical Society, the Pennsylvania State Archives, the Philadelphia City Archives, the Springfield Township Historical Society, and the Thomas Cooper Library at the University of South Carolina.

Beyond the general staff of these institutions, however, there were numerous saints who went beyond the call of duty, either in providing access to materials not normally available to researchers or in working with me to secure rare images and permission to use those images in this book. I am especially grateful to Richard Boardman, department head of the Map Collection at the Free Library of Philadelphia, who spent the better parts of two days allowing me to rifle through his many, many maps and atlases, then listened patiently and attentively as I explained exactly what was needed to show Erdenheim's shifting property lines across three centuries. I am also deeply indebted to Cathy Schenck, librarian at the Keeneland Library in Lexington, Kentucky, who went down the rabbit hole with me for three days to dig up images of horses, racing silks, and owners from 150 years ago, then plugged some holes for me from a distance when I still had questions. Nicole Joniec, an assistant in the Print Department at the Library Company of Philadelphia, went into superfast mode and helped me secure an image I discovered desperately late in the game, while Hillary S. Kativa, the rights and reproductions specialist at the Historical Society of Pennsylvania, helped me find three essential images I could not find on my own at the last possible moment. Lionel Green, the records supervisor at the Montgomery County Archives, found the perfect balance of letting me fend for myself and stepping in to save me from myself when I needed it, all the while replenishing reams of paper and several toner cartridges as I printed out scores of wills, deeds, and tax records. Jonathan Stayer of the Pennsylvania State Archives in Harrisburg graciously sat with me and walked me through the labyrinth of Pennsylvania's land records, allowing me to crack the code of patents, warrants, and surveys. I am also immensely grateful to Jim M. Duffin of the Penn Archives at the University of Pennsylvania, not only for his landmark work in creating the *Mapping West Philadelphia: Landowners in October 1777* online project, but also for his generosity in creating a customized map of some of this data for me that I could use explicitly for this project. Without his work and his assistance, there would be no easy visual demonstration of what Erdenheim's boundaries looked like during the colonial period or the early nineteenth century.

In addition to the institutions and their employees mentioned above, I am also grateful to Jackie Penny at the American Antiquarian Society, Teresa O'Toole at the Sterling and Francine Clark Art Institute, Heather Burychka at the Springfield Township Historical Society, William L. Gross at Carson Valley Children's Aid, the Copy Services staff at the Minnesota Historical Society, Michelle Clark at the Frederick Law Olmsted National Historic Site, Margery N. Sly at the Special Collections Research Center for the Temple University Libraries, and Rebecca Howard at the Harness Racing Museum and Hall of Fame, all of whom helped me track down hard-to-find historical images, get them reproduced, and secure permissions so that they could be used in this history. Kevin Archie, my old friend and fellow photographer from Columbia, South Carolina, also stepped in on short notice with gorgeous representations of the Kittson and Widener racing silks when my Photoshop skills proved insufficient for the task. Cynthia Colbert, professor of art at the University of South Carolina, prepared a stunning set of handmade boards with a handmade binding for a special edition of this book given to Peter and Bonnie McCausland. Perhaps most importantly, I am deeply indebted to the staff of the University of South Carolina Press for their patience, understanding, and hard work on this edition of the book, especially Suzanne Axland, Elizabeth Jones, Pat Callahan, and Jonathan Haupt.

My real heroes, though, are the folks who helped when they weren't being compensated and gave to this project simply because they were fascinated by it. Maryanne Buck of the Morris Arboretum may not have had any materials there that could have helped me with the project, but she was kind enough to do a little research on her own and then point me to an essential resource I never would have found without her help. Denise V. Berg, author of *The Welch Resort,* a book about the children of Aristides Welch, didn't have much she could offer directly to me about the Welches, but she put me in touch with someone who could. Her contact, Mary Jane Echenhofer, Aristides Welch's great-granddaughter, proved to be a gold mine of long-hidden information about the Welches, and I look forward to our many conversations in the coming months as we compare notes about whom I believe to be Erdenheim's most fascinating, and maddening, owner. And I would be terribly remiss if I did not thank Frank Garber, proprietor of the Chestnut Hill Camera Shop at 8614 Germantown Avenue, who kindly reopened his shop's doors at 5:30 P.M. on a Saturday afternoon (well after closing time) and allowed me with my 103-degree fever and desperate pleas to purchase replacement batteries for my old-school Bronica ETRSI-II film camera just before I went trespassing on the grounds at Erdenheim. It was a glorious late afternoon in mid-May, the perfect slanted light of the Golden Hour just emerging, and Frank made sure I didn't miss it.

Beyond all of these, I am most grateful for the people closest to me who endured this project and all I had to say about it. My seventeen-year-old son, Griffin, who shows all the signs of having inherited my incurable "love of history" disease, willingly and (mostly) gleefully traveled with me on one of my excursions to Philadelphia, took notes for me and compared notes with me, helped me solve the mystery of where "Old" Erdenheim disappeared to, waited on a bench in the lobby of the Library of Congress when he was told he was too young to enter, and otherwise served as an outstanding copilot and confidante. My gorgeous and brilliant wife, Teresa, meanwhile, listened patiently as I told her all about my Erdenheim journeys and discoveries, every day, sometimes in my sleep; asked all the right questions and made all the right suggestions about how to solve dozens of research problems; and never once asked whether her husband was ever going to get off the road and come back home again. I am a lucky, lucky man.

Finally I want to thank Dean Fitzpatrick and Dr. Glickman—and indirectly Peter and Bonnie McCausland—for this unexpected and rare opportunity to write about Erdenheim and its fascinating past, even if the entire experience was something of a whirlwind. If I had another week, or month, or year to complete this history, I'm certain I could tease out even more from the records and sources that survive, including things I have not yet found, and thus help some of Erdenheim's owners and residents come alive in a more palpable way. Alas, this is the regret of every historian. All projects must necessarily come to their end. Along the way, though, I learned a lot, especially about things that I didn't know much about before—horse racing, farming, hunt clubs, early streetcar companies, colonial and antebellum life on the outskirts of Philadelphia, and, of all things, Sirhan Sirhan. I also had far more fun than I ever expected I would. And after all, those two standards—education and enjoyment—are the true measure of the value of a history, whether one happens to be reading it or writing it. Telling Erdenheim's story has filled me with both.

NOTES

1. William F. Buck, "Whitemarsh Township," in Theodore W. Bean, ed., *History of Montgomery County, Pennsylvania,* Philadelphia: Everts and Peck, 1884, 1138.
2. Aziz Shihab, *Sirhan* (San Antonio: Naylor, 1969).
3. The entire story of Sirhan's life in California is gleaned from Shihab, *Sirhan,* 71–76.

CHAPTER ONE

The Days of Yeoman Farmers

What we know of the land that would become Erdenheim Farm, in the time before William Penn started parceling out land grants to European settlers eager to start new farms, mills, and other opportunities in the Pennsylvania colony, is somewhat scant. The warrant and survey for the earliest land grant for the Whitemarsh portion of what would one day be Erdenheim—a 5,000-acre grant to Jasper Farmar Jr.; his father, Major Jasper Farmar; and his brother Richard Farmar made in January 1683 (O.S.)—makes clear that this was the land "ye Indians call'd Umbilicamence."[1] That same year Thomas Fairman, traveling with one of the Proprietors and several other men, reported having been to "Umbolekemensin," a Native American community that nineteenth-century historian Theodore Bean placed unmistakably "within the rich valley of the Wissahickon and not more than a mile from the centre of the township [Whitemarsh]."[2]

Those "Indians," the oldest sources tell us, were the Lenni Lenape, one of the Five Nations of Native American tribes better known as the Iroquois. The Lenni Lenape, meanwhile, came to be known to the European settlers as the Delaware, and their territory included almost the entirety of the land between the Hudson and Susquehanna Rivers. Upon his arrival in 1682, William Penn was wise enough to conclude an immediate treaty with the Delaware, one structured on a set of agreements that allowed relations with the Native Americans in Pennsylvania to proceed much more peacefully over the ensuing seventy years than they did in most of the other American colonies. Penn's agents and the Lenni Lenape concluded the key treaty relative to the land that would become Erdenheim on July 14, 1683, when the native tribes agreed to sell all the lands between the Schuylkill River and the Pennepack, as far north as the Conshohocken hills, thus conveying a portion of Whitemarsh. A second treaty on July 30, 1685, conveyed the remainder of what would become Montgomery County to the white settlers, except for a portion lying east of Pennepack Creek. The price, as indicated in the treaty, was not particularly steep: "200 fathoms of wampum, 30 fathoms of duffells, 30 guns, 60 fathoms of strawd waters, 30 kettles, 30 shirts, 20 gun belts, 12 pairs shoes, 30 pairs stockings, 30 pairs scissors, 30 combs, 30 axes, 30 knives, 21 tobacco tongs, 30 bars of lead, 30 pounds powder, 30 awls, 30 glasses, 30 tobacco boxes, 30 papers of beads, 44 pounds of red lead, 30 pairs of hawks' bells, 6 drawing knives, 6 caps, [and] 12 hoes."[3]

FACING: Erdenheim cattle at sunset, Angus Tract, May 2012. Photo by Eric Plaag.

Reproduction of Thomas Holme, *A Map of Ye Improved Part of Pennsilvania in America, Divided into Countyes, Tonwships and Lotts . . .*, 1687, reprint from appendix, *Pennsylvania Archives,* Third Series, Map Collection, Pennsylvania Historical and Museum Commission, Pennsylvania State Archives. The Wissahickon is indicated as "Whitpaine Creek" on the Holme map.

With their land patent in hand, the Farmar family arrived from England on the *Bristol Merchant* at Philadelphia on December 10, 1685, intent on establishing a new community on their land called "Farmar Town." Shortly before their departure, though, Major Farmar died, and in the months after their arrival, his son, Jasper Farmar Jr., died, too, leaving the family estate in a bit of turmoil. Through a series of transactions intended to divide the land holdings according to Major Farmar's will, Major Farmar's widow and her son Edward became the joint holders of 3,750 acres of the original land, with Edward inheriting the entirety of this estate upon his mother's death in 1686. Catharine Farmar, the widow of Jasper Farmar Jr., retained the other 1,250 acres located in the northwest corner of the original parcel (well outside the boundaries of present-day Erdenheim).[4]

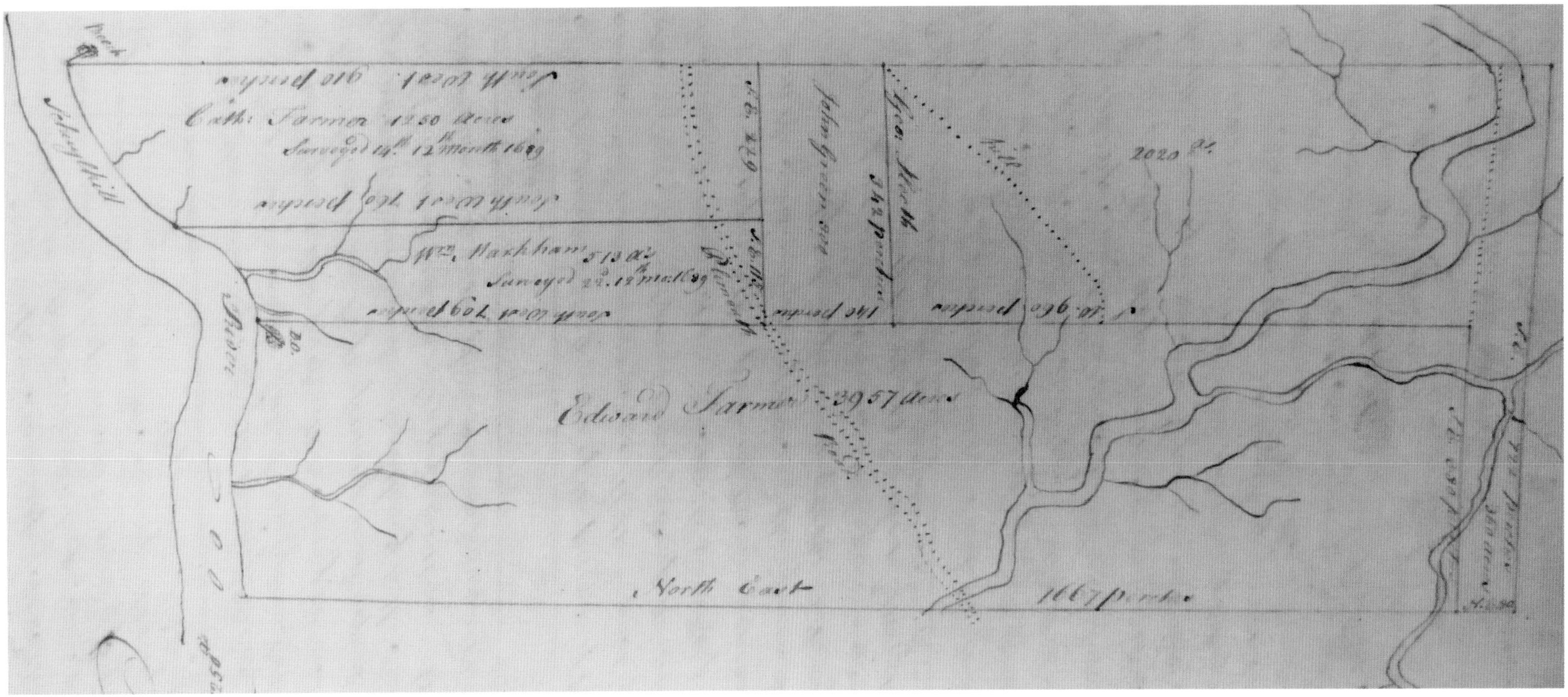

Survey of Farmar family lands in Whitemarsh, 1703. Philadelphia City Archives.

In the years that followed, Edward Farmar established an outstanding reputation as an interpreter to the Native Americans, as a Philadelphia County court justice for nearly forty years, and as a member of the colonial Assembly for thirty-five years, having been elected to that body in 1710. His mill at Whitemarsh (present-day Flourtown), meanwhile, was instrumental in producing much of the lime used by settlers in this region during the late seventeenth and early eighteenth centuries. This lime was produced from the ample bed of limestone that ran directly beneath Edward Farmar's parcel and the land that would one day be Erdenheim. So highly regarded was the limestone produced by this seam during Madame Farmar's brief maintenance of the estate that a representative of the Free Society of Traders in Pennsylvania advised William Penn that her limestone was available for sale and was "as good lime-stone . . . as any in the world . . . and near to your [Manor] of Springfield."[5]

Penn's Manor of Springfield, as indicated on the famous Thomas Holme map of the region, lay directly to the southeast of Edward Farmar's principal Whitemarsh parcel. The original warrant and survey for this property originated on the August 1682 orders of William Penn himself, who called for a survey to lay out "Five Thousand Acres of Land for a Mannor for myself called Springfield according to Direction I gave to Thomas Fairman."[6] Penn named the manor "Gulielma Maria Penn's Manor of Springfield," in honor of his wife, allegedly so that she and her descendants could forever reach the Schuylkill by their own land.[7] Soon after her death, however, Penn began selling off parcels within the manor, so that by 1732, Springfield Manor was subdivided into seventeen different parcels. Edward Farmar owned four of these widely scattered throughout the manor, including a 247-acre parcel immediately southeast of the Whitemarsh Township line encompassing the portion of present-day Erdenheim that lies within Springfield Township.[8]

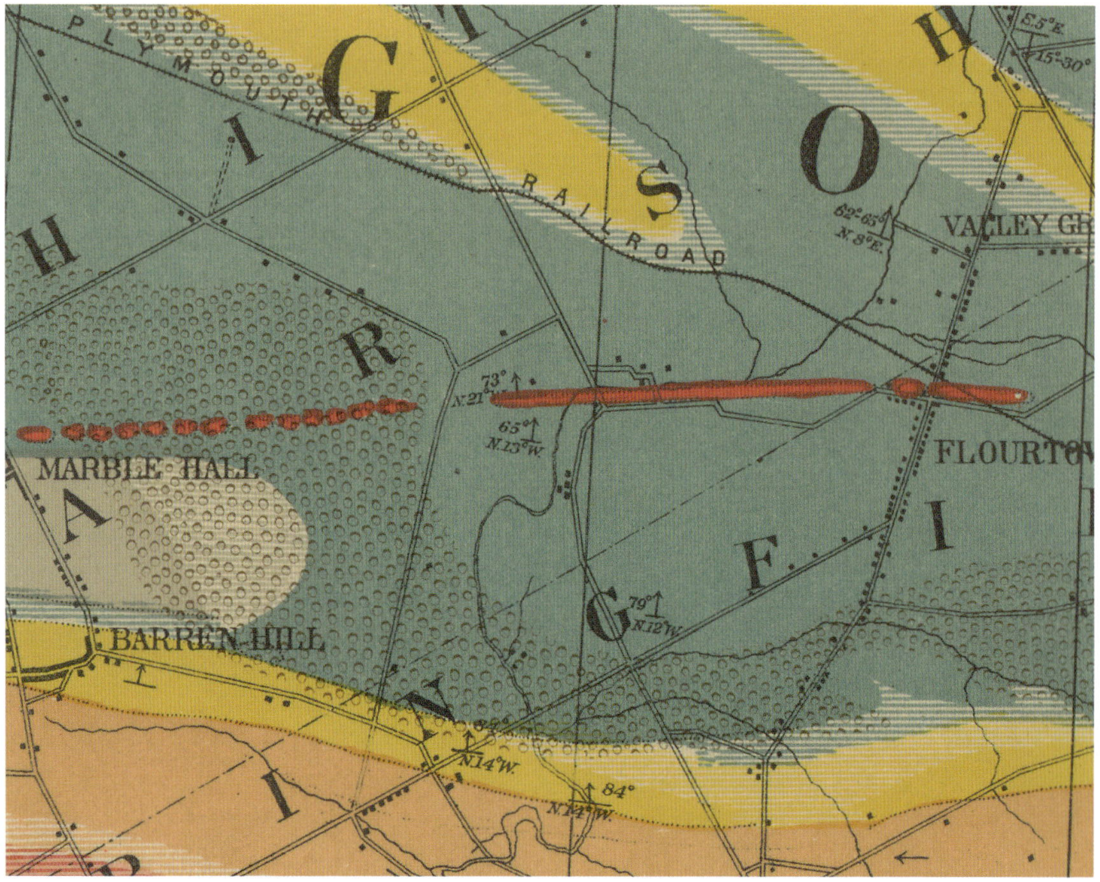

Detail of Erdenheim Farm area from 1895 *Geological Map of the Philadelphia Belt of Azoic Rocks,* plate 11260-1, Map Collection, Free Library of Philadelphia. The turquoise seam is magnesian granite and limestone.

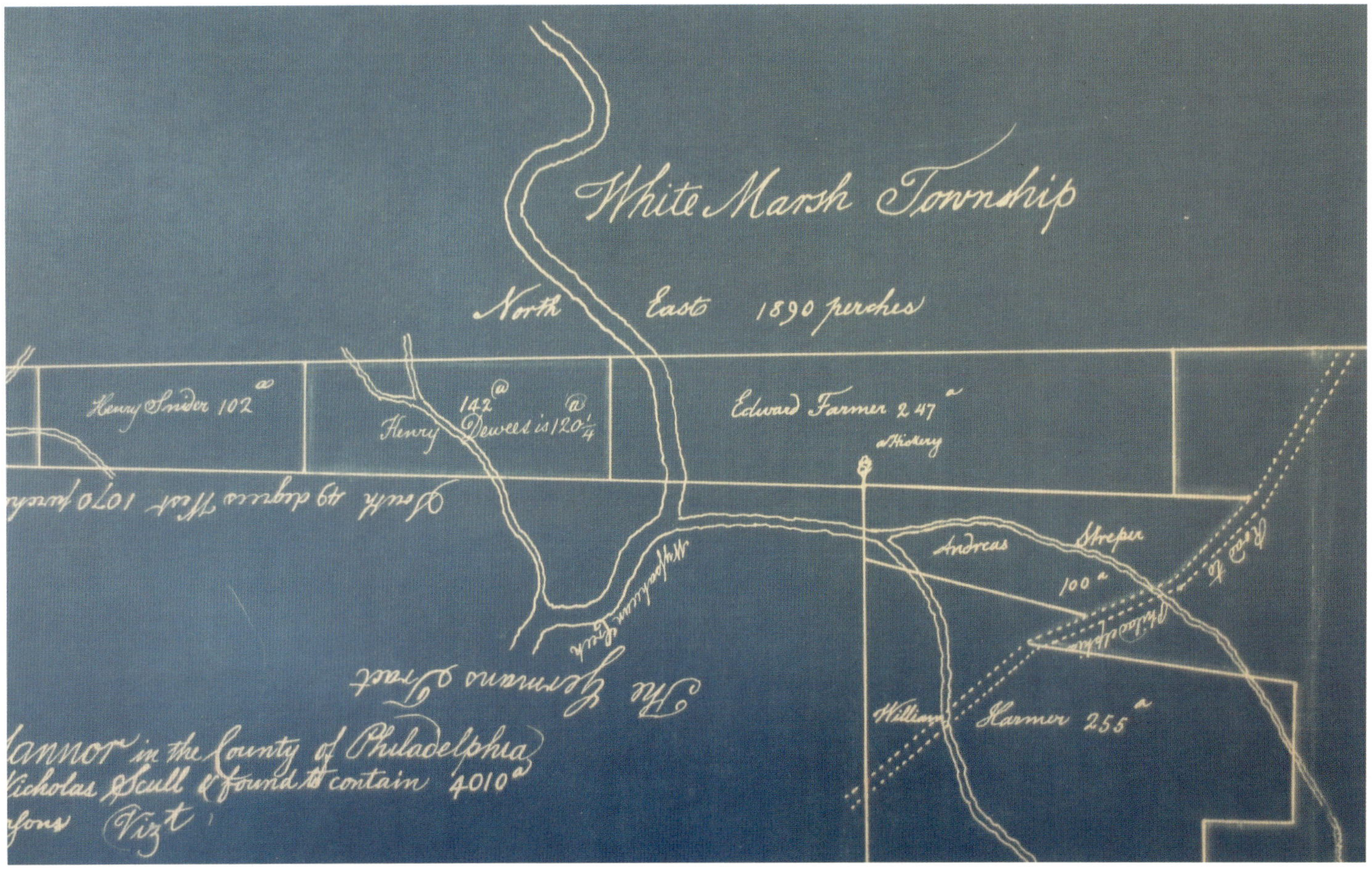

Detail from Nicholas Scull's 1732 Map of Springfield Manor, Land Office Map Collection, 17.522, Map #3724, Pennsylvania Historical and Museum Commission, Pennsylvania State Archives.

Peter Robeson, Jonathan Robeson, and Their Land along the Wissahickon

Upon the death of Edward Farmar on November 3, 1745, at the age of seventy-three, settlement of his vast estate fell to his son Joseph and to Peter Robeson and the latter's nephew Jonathan Robeson, each of whom had married Edward Farmar's daughters, Sarah and Catharine, respectively. Through these connections both Robesons came into substantial landholdings in Whitemarsh, including portions of the land that would one day be Erdenheim. By 1742, however, Jonathan Robeson already possessed much of the future Erdenheim territory, totaling 377 acres entirely within Whitemarsh.[9]

Seal of Edward Farmar, from Bean's *History of Montgomery County.*

Little survives to tell us much about Jonathan Robeson and his use of this land, but there are a few indications of his reputation and position in colonial Pennsylvania. In addition to his connections to Farmar, Jonathan Robeson played a pivotal role in protecting the lands of the proprietors in Springfield Manor. A 1727 ranger appointment by James Logan, William Penn's former secretary and a member of the Pennsylvania Provincial Council, called on Robeson "to take care of the Timber and wood [——]upon his Mannor of Springfield, and strictly forbid all persons who shall perform to cut down or destroy the same."[10]

As for the rest of the Farmar lands relevant to Erdenheim, Jonathan and Peter Robeson, as well as Joseph Farmar—all acting as executors—sold a total of 952 acres to Thomas Yorke and his wife, both of Germantown, on October 19, 1747. This parcel, according to the region's oldest historians, included much of the land to the east of present-day Stenton Avenue and north of the Wissahickon as far as present-day Joshua Road, as well as the parcel south of the Wissahickon between present-day West Mill Road and the Springfield Township line, and much of the land on the west side of present-day Stenton Avenue included in Erdenheim's current boundaries. The following day Yorke transferred the entirety of this 952-acre parcel to Peter Robeson, suggesting that the entire set of transactions was a complex maneuver to facilitate legal title in light of Robeson's restrictions on direct purchase as one of the executors of Farmar's estate.[11]

Sometime after taking possession of this tract, either Peter or Jonathan Robeson built a mill directly on the Wissahickon on the north side of the intersection of present-day Stenton Avenue and West Mill Road. Known in later years as the Cleaver Mill, the mill remained in operation into the late nineteenth century.[12] This industrial concern aside, the Robesons did not hold on to their original 952-acre parcel for very long. On March 4, 1755, Peter Robeson divested himself of 341 acres at the lower or southeast end of the original parcel by selling that land for 1,483 pounds to Nathan Sheppard of Abington Township.[13] That same day, Robeson sold 555 acres for just over 1,666 pounds to Anthony Williams Sr., of Bristol Township, husband of Sarah Shoemaker, this transaction representing the more northwesterly portion of the original Yorke parcel. This land would have included much of the present-day Erdenheim lands on both sides of Flourtown Road between Stenton Avenue and Thomas Road.[14]

The Old Burial Ground

Nearly simultaneous to these transactions, on January 8, 1744, Edward Farmar sold to his eldest son, Samuel, 24.75 acres of land located between the Wissahickon and present-day Stenton Avenue. Two years later Samuel Farmar and his wife, Mary, sold one half acre of this parcel to Henry Bartleson, Peter Knight, and James Stroud "in Company" for the small sum of five pounds, said land to be used exclusively for the establishment of a graveyard. Over the next eighty years, elegant headstones marked the graves of some of those who were buried within its "substantial wall . . . of considerable size."[15] On August 2, 1786, Peter Knight—at that time the only surviving trustee from the original deed—made a declaration of trust to George Hocker, Nicholas Kline, and others, placing them and their heirs and successors in control of the cemetery. In addition to emphasizing that the land must be permanently used for a burial ground, it granted the trustees the right to build "a House of Worship thereon" (which was probably never built) and ordered them to transfer trusteeship "when the Sd. Trustees, by age or infirmity become unable to perform the duties of the Sd. trust."[16]

The headstones and perimeter wall of the cemetery appear to have survived until about 1916. William A. Yeakle described the cemetery in 1895 as lying "on the property owned by Lewis A. Lukens, of Conshohocken, adjoining the Williams school."[17] Forty-five years later, William R. Yeakle remembered that it lay "just above a southern slope, close by the location of the original Erdenheim, and at a spot which commands a view of the lower Whitemarsh valley." A good estimate of its more exact location is that it lay directly opposite the current home known as Erdenheim—what is called "New" Erdenheim in this history—on the ridge midway between and parallel with the private "New" Erdenheim driveway leading to the stone bridge across the Wissahickon.[18] Describing the burial ground further, the younger Yeakle wrote,

> I can remember when, as a pupil at the Williams school which stood about three hundred yards distant, being in and around the burying-ground many times and reading the inscriptions on the stones. I remember at that time (1904–1908), the stone wall had partially fallen at several points and I have a recollection of an iron gate swung open against the wall. The enclosure

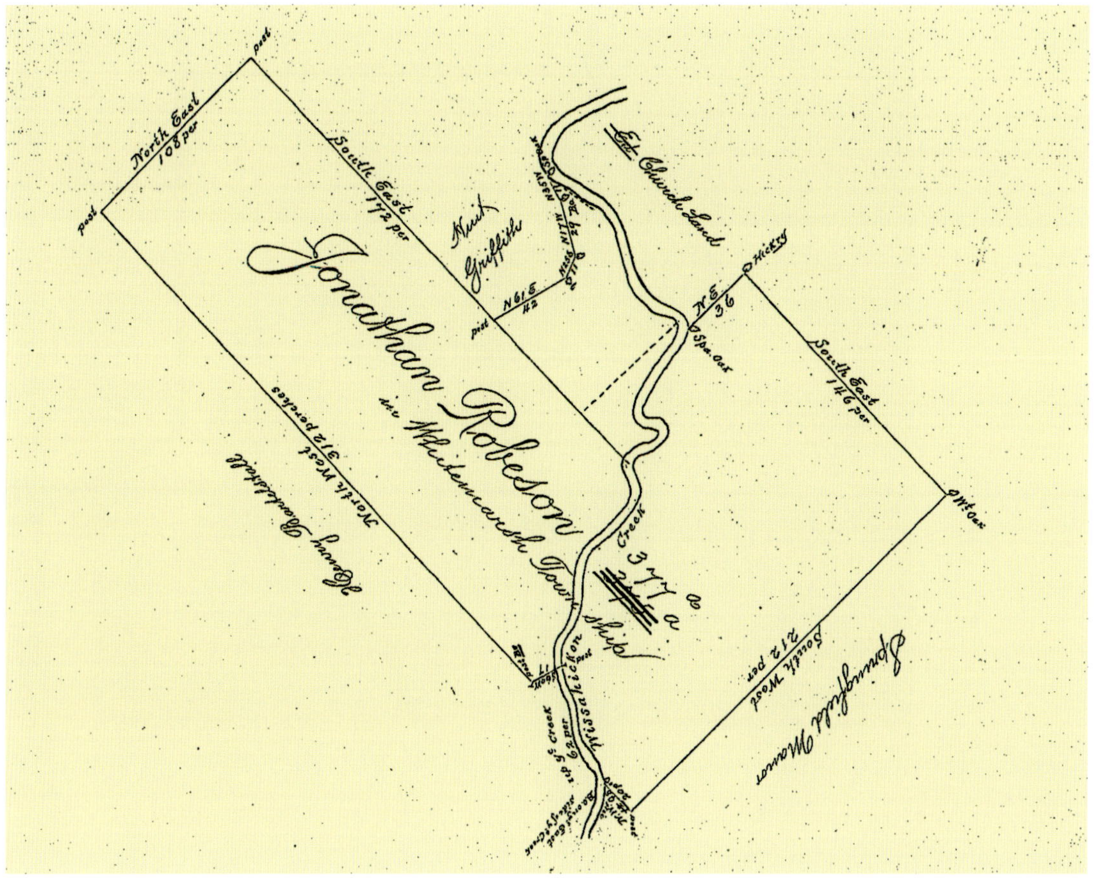

November 17, 1742, survey of Jonathan Robeson's land in Whitemarsh, Copied Surveys, Series 17.114, Record Group 17, Records of the Land Office, book B7, 206, Pennsylvania Historical and Museum Commission, Pennsylvania State Archives. Digital map overlays suggest that the lower portion of the parcel in question would have extended from where the Wissahickon crosses the Springfield-Whitemarsh line northeast to a point just to the east of present-day Stenton Avenue. The upper portion would have extended northwest to a line near the intersection of Flourtown and Thomas Roads.

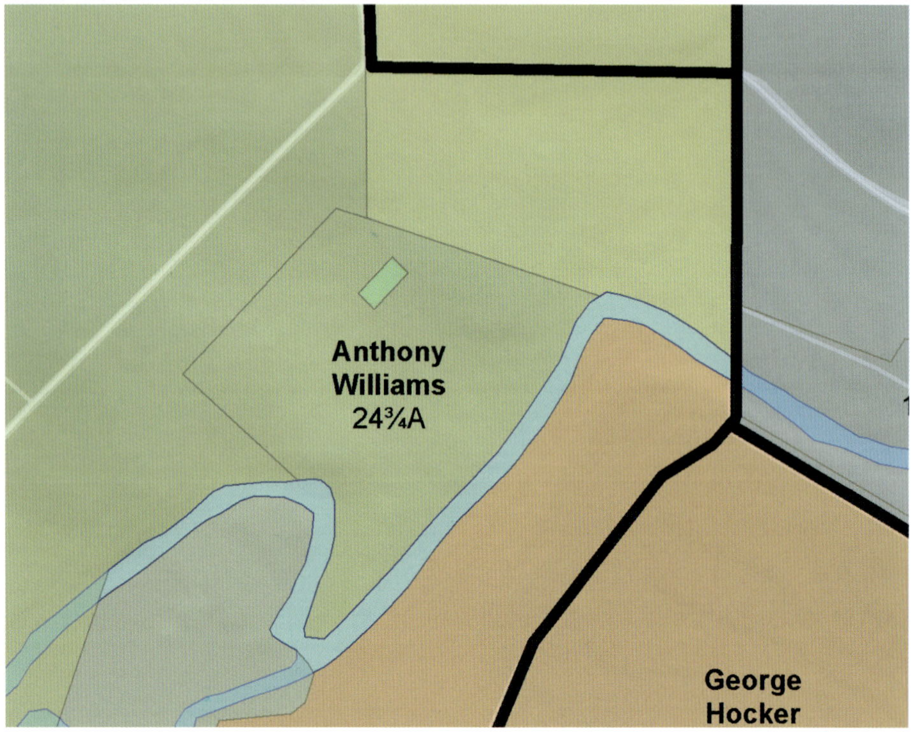

Detail from J. M. Duffin, *Mapping West Philadelphia: Landowners in October 1777*, University of Pennsylvania Archives. The cemetery parcel is the lime green rectangle near Anthony Williams's name.

> contained many marked graves and several of the stones seemed to be unusually large. One, in particular, stood out from the rest. It was larger and more elaborate and bore certain symbols, carved in bold relief, around the top. . . . During the time I attended Williams school and a number of years previous, the surrounding ground was owned by Robert N. Carson and the school-children were allowed to play over the property at will. Sheep grazed over the land and also kept the grass closely cropped in and about the grave-yard.[19]

In his history of Whitemarsh, William A. Yeakle offered a listing of those burials marked by tombstones at the time of his writing. His list is as follows:

> John Nichol Knight, died December 29, 1722, aged 40 yrs., 10 mos.
> Susanne Menan (child), died February 20, 1763.
> Susanne Menan, died October 28, 1787, aged 71 years.
> Sarah Menan, died December 12, 1796, aged 39 years.
> John, son of Patrick and Susanne Menan, died at Penn's Neck, October 12, 1768, aged 25 years.
> Patrick Menan, died February 5, 1791, aged 80 years.
> Levi Trump, died November 8, 1784, aged 29 months.
> Mary, wife of Jacob Siddon, Died May 24, 1797, aged 26 years.
> Bartle Bartleson, died February 17, 1777, aged 80 years.
> Elizabeth Bartleson, died March 24, 1769, aged 60 years.
> Cephas Bartleson, died September 15, 1783, in his 56th year.
> Elizabeth Bartleson, died August 13, 1783, aged 43 years.
> Susanne Bartleson, aged 1 year, 3 months.
> George Bartleson. Aged 9 months.
> Margaret Heinch, died November 11, 1763 or 5.
> Nicholas Knight, died November 7, 1787, aged 69 years.
> J. K. (probably Knight), died 1758.
> Elizabeth Coleman, died August 29, 1746, aged 68 years.
> Catherine Coleman, died April 11, 1746, aged 3 years.
> Elizabeth Coleman, died April 26, 1746, aged 19 months.
> Christopher Mason, died December 3, 1780, aged 46 years.

Ann Mason, died May 6, 1802, in her 75th year.

Peter Mason, died October 30, 1804, aged 38 years.

Charles Bilger, died August 9, 1821, aged 6 years.[20]

The younger Yeakle's account, meanwhile, reveals what became of this old burying ground containing the remains of some of Whitemarsh Township's earliest residents, but for that story, we must wait until Erdenheim's history reaches the twentieth century.

Johann Georg "George" Hocker and Erdenheim

As with many of Whitemarsh's landowners during the colonial period, Nathan Sheppard soon grew weary of his Whitemarsh parcel, selling 333.5 of the original 341 acres to William Streper—a well-known landowner in both Germantown and the surrounding townships—on April 11, 1761.[21] It was in this same year that Streper built a gristmill directly on the Wissahickon—a mill building and water wheel located just a hair downstream from Erdenheim's present-day southern boundary, both of which were replaced in 1854 by John Piper.[22] Just two years later, in 1763, Streper—who seemed to buy and sell properties in the vicinity with as much of a sense of speculation as the Farmars and the Robesons before him—sold just over 200 acres of the Sheppard parcel to a German immigrant, Johann Georg Hocker, better known as "George" to both his neighbors and the courts. Concluded on April 21, 1763, the sale was a complicated transaction that reversed an intended sale of a portion of this tract to a William Antes for the sum of 1,600 pounds paid by Hocker. In this conveyance Streper retained the right to raise and lower the water levels of the Wissahickon on Hocker's parcel as he saw fit, presumably because of the presence of a mill on the land Streper retained south of there along the Wissahickon.[23]

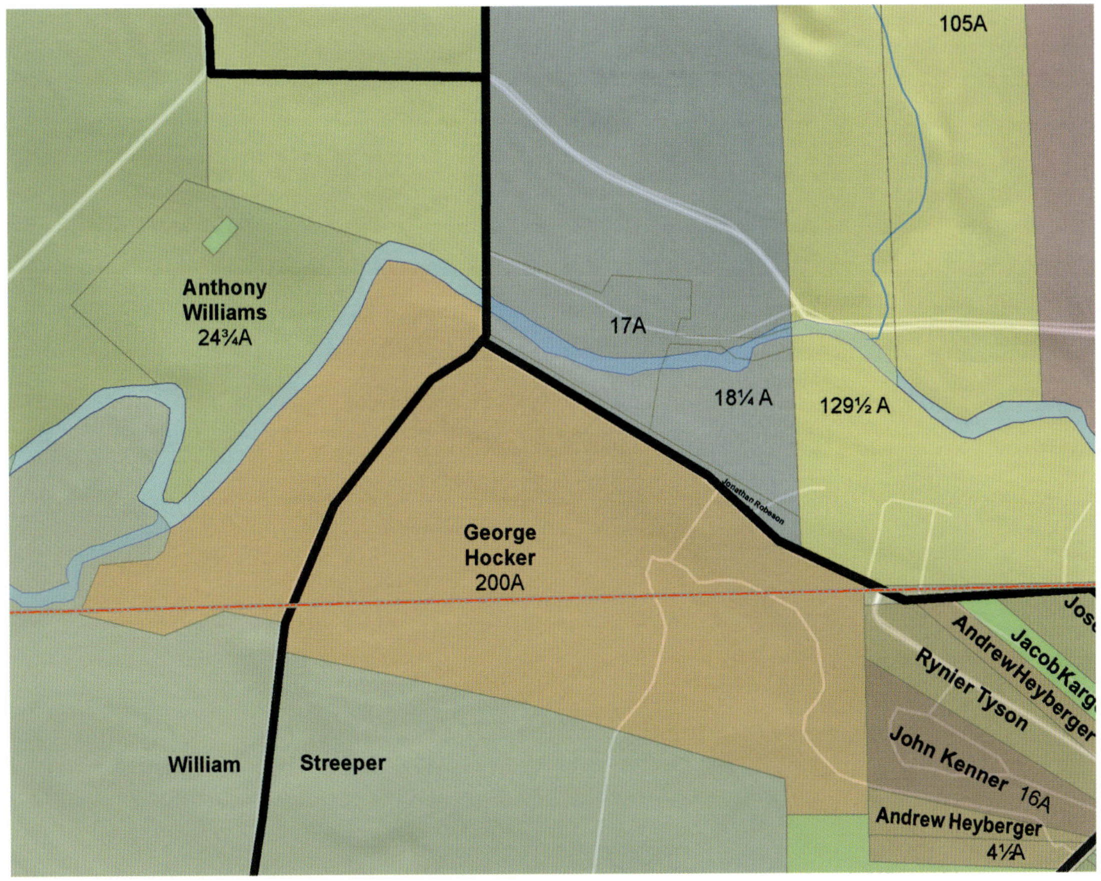

George Hocker's 200-acre parcel in 1777, from J. M. Duffin, *Mapping West Philadelphia: Landowners in October 1777*, University of Pennsylvania Archives.

The original or "Old" Erdenheim, as it appeared circa 1922. This image was apparently taken by local historian Edward W. Hocker and given to Samuel Yeakle shortly before Yeakle referenced it in his April 1922 address on Erdenheim. The same image appears in the 1935 published version of this address. Image from the collection of the Springfield Township Historical Society.

Born in Wurtemberg, Hocker arrived in Philadelphia in September 1751 at the age of eighteen. He later married Margaretta Weidmann in Lancaster County, then naturalized in Lancaster in 1760. It is not entirely clear whether Hocker built the original home that came to be known as "Erdenheim," or whether the house already stood there at the time he purchased the land. Tradition holds that Hocker named the place to mean "earthly home," and one later source suggests a construction date of 1764, although our story will eventually reveal that these details, too, may be spurious.[24] William A. Yeakle dates the house to sometime before the Revolutionary War and asserts that it served as a tavern for many years "in the olden time." Samuel Yeakle, speaking before the Historical Society of Montgomery County in 1922, asserted that the name Erdenheim itself dated to the Revolution.[25]

To a modern observer unfamiliar with the topography of Erdenheim in Hocker's day, the historic position of "Old" Erdenheim on the present-day farm might seem counterintuitive, but in fact the old house aligned smartly with the local infrastructure and the conventional domestic priorities of its day. Located directly adjacent to the Wissahickon on its left or southeast bank, "Old" Erdenheim looked downstream toward due south. An old eighteenth-century road also forded the Wissahickon just below the house, then joined the Lancasterville Road (Stenton Avenue) near the current main entrance to Erdenheim on Stenton's west side. Such a position gave the residents of "Old" Erdenheim direct access to the road and the water and ensured a good deal of sun for the living quarters located at the front of the house.

As it turns out Hocker may have nearly lost the Erdenheim property, a portion of which he apparently mortgaged for 1,700 pounds in April 1768. By then the property may have also contained a mill. Hocker successfully paid off a portion of this debt at various times, but the interest accruing eventually left

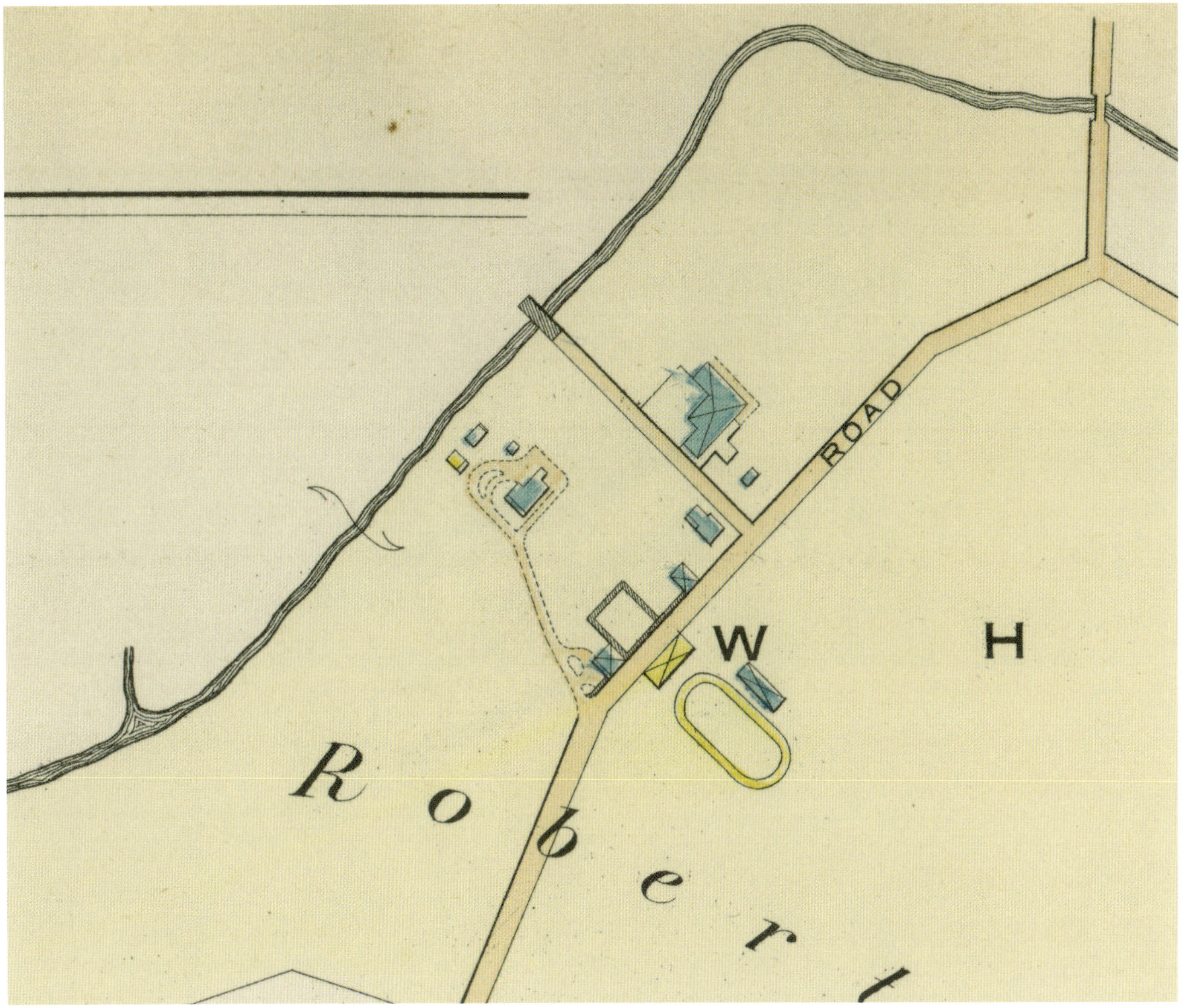

Detail from Ellis Kiser, Otto Barthel, and B. A. Skeels, *Atlas of Cheltenham, Abington, and Springfield Townships and Vicinity, Montgomery County, Pa.*, 1897, plate 18, Map Collection, Free Library of Philadelphia. "Old" Erdenheim was the house indicated in blue and located within the grand piano-shaped driveway complex just above the "o" in "Robert." The ford across the Wissahickon for the eighteenth-century road was located at the double-arrowed wave just downstream from the house.

him even more deeply in debt, prompting Hocker to sell the mill and part of the land to "Weiss" for 1,750 pounds sometime prior to April 1777, the entirety of which was to be paid to the mortgage holder. The ensuing confusion over the payment of this debt left Hocker at the center of one of the earliest cases to reach the Pennsylvania Supreme Court in 1789. In the case of *Johnson v. Hocker,* 1 U.S. 406, the plaintiff—a "Mr. F. Johnson of Germantown"—alleged that Hocker had taken out a bond on April 24, 1777, for 500 pounds on the remaining debt and had never paid it. Hocker countered that he attempted to make payment and called on a certificate from the local treasurer to be read to the jury, in which the treasurer confirmed the payment of two thirds of the bond amount in the presence of witnesses, but noted that the plaintiff had refused the payment, perhaps because Hocker had attempted to pay in deeply depreciated colonial currency. Hocker, the treasurer's certificate indicated, then gave the remaining third to the treasurer, Isaac Snowden, and instructed him to give that sum "to such poor and distressed persons as I shall think proper objects of charity."[26]

The plaintiff's attorney objected to the introduction of the certificate, indicating that its contents were the equivalent of testimony and injurious to his client, and as a result, that the plaintiff had the right to cross-examine Snowden. Hocker's counsel countered that the superfluous information could be excluded from the jury's view, with only the certificate as evidence of payment being provided. The chief justice thus read only those portions relevant to the payment of the debt to the jury.

In considering the case, the Pennsylvania Supreme Court concluded that Hocker had indeed attempted to pay his debt, but it looked carefully at the *manner* in which he had paid it, noting that if it had been paid in bills of credit issued prior to January 29, 1777 (the date when new laws governing payment of debts were passed in Pennsylvania), then the discharge of the debt was absolute simply by the act of payment. If it was paid in specie issued after January 29, 1777, then per Pennsylvania law, the debt should be "liquidated according to the scale of depreciation" on the currency used. Thus, the court concluded, Johnson was entitled to receive "only so much as that sum [500 pounds] in paper money was worth at the time the contract was entered into, . . . nor is he entitled to any interest from the date" when Hocker attempted to pay the debt. Using this calculation the court concluded that Hocker still owed Johnson just over 272 pounds on the old debt, less that portion of which was paid in bills of credit issued prior to January 29, 1777.[27]

While it is not entirely clear that this debt centered around the Erdenheim lands, what is clear is that by 1785, George Hocker was taxed on only 70 acres of land, his dwelling, four horses, and three cows, nearly 38 acres of which must have been the land he acquired in two separate transactions from William and Margaret Streeper in August 1784 (inclusive of a mill race) and John Streeper in January 1785. As described in the deeds, these lands appear to have been a part of the old Hocker lands northwest of present-day Northwestern Avenue, southeast of the Wissahickon, and southwest of Stenton Avenue.[28] Over the next year, Hocker added more real estate to his holdings, expanding to 137 acres by the time he was taxed in 1786, although he may have divested himself of 7 acres of this land—or leased it out—between 1793 and 1798, when he was taxed on only 130 acres.[29]

Whatever his financial woes might have been during the late eighteenth century, Hocker also seems to have been plagued by wandering horses and cows, who no doubt found the tasty, placid waters of the Wissahickon and the relatively low banks on Hocker's land appealing. Hocker placed numerous ads during the late eighteenth century in the *Pennsylvania Gazette,* alerting neighbors to the presence of strange livestock on his farm. In each case Hocker invited the owner to "come, prove property, pay charges, and take her away." Perhaps the most interesting of these advertisements is the one that appeared on June 8, 1791, clearly listing Hocker as an innkeeper, thus confirming the persistent legend that "Old" Erdenheim was—at least at that time—an inn.[30]

What else we know of George Hocker is scant indeed, although family lore records that he "was the discoverer of the value of gypsum as a fertilizer for grass" and accidentally discovered its use when storing old ship ballast on the fescue near his mill.[31] Hocker also appears to have been the occasional target of criminals, perhaps because of his home's location just off two significant roads through the region and near the only reliable ford at that time on that stretch of the Wissahickon. William A. Yeakle, for example, relates the story of a vicious attack that occurred at Hocker's farmhouse around 1800, shortly after Hocker had sold some property. Because it was widely known that he still had the cash at the house, Yeakle suggests, Hocker's attackers plotted a robbery. Yeakle continues,

> About midnight he was called out of bed (the family had all retired) by [the robbers] for the purpose of warming themselves. This, on account of the unusual hour of the night, was denied them. Owing to their persistency, he called up his hired man, who came at once to his assistance; but by [that] time the burglars had broken open the door and effected an entrance, and demanded his money, which he at once denied them. Mr. Hocker, wife, and hired man, the only ones at home, except a small daughter, attempted to drive them from the house. The ruffians, believing themselves strong enough, resisted them, and a bloody fight at once commenced. There being two burglars, the hired man attacked one of them and Mr. Hocker and his wife the other. After fighting some time, Hocker, being an old man, and nearly exhausted, made a desperate stroke at the burglar, who had temporarily left him in pursuit of his wife with a drawn knife, and felled the villain to the floor. . . . The old man, thus relieved of his antagonist, went to assist the hired man, who was getting the worst of it, and their united efforts soon placed him hors du combat. They at once secured him with ropes, etc., and then treated the other in the same manner. No life was lost, but blood flowed quite freely, as the walls, entry, furniture, floors, etc., bore evidence of this.[32]

Hocker was then assisted by guests leaving an apple-dunking party at Streper's home nearby, who took the burglars to Norristown; from here, the robbers escaped from jail, then were recaptured, "tried, convicted, and sentenced for a long time to hard labor."[33]

When George Hocker died a widower on October 16, 1820, he left no will, throwing the disposition of his estate into a bit of chaos, given the five surviving children, the surviving heirs of his three deceased children, and dozens of descendants scattered about Pennsylvania with a keen interest in the estate's future. At his death the Erdenheim lands comprised 163 acres, 130 perches, and left with little idea of how to resolve the estate, George Hocker's son Martin Hocker and son-in-law Henry Daub petitioned the Orphan's Court of Montgomery County for resolution in November 1821.[34] The court determined that the Hocker estate could not be partitioned equally, and the Hocker heirs determined that they would not accept the valuation of $150 per acre assigned by the court in February 1822, thus exposing the entire property to public auction. On August 17, 1822, Martin Hocker, who had been named administrator of the estate, sold 155 acres, 110 perches of the George Hocker parcel at a valuation of $100 per acre to Casper Schlater of Upper Dublin Township. In January 1823 Martin Hocker also filed his estimation of the value of the estate, placing a value on the personal inventory of just over $9,151. Sadly, no copy of the personal property inventory survives.[35]

The Williams Parcels

As a result of the 1755 division of the Peter Robeson parcel, Anthony Williams Sr. possessed 555 acres of land in Whitemarsh, much of which would eventually be part of present-day Erdenheim. Fifteen years later Williams acquired another key parcel to Erdenheim's history—the 24.75-acre plot originally conveyed to Samuel Farmar in 1744, within which lay the old half-acre burial ground plot. Located along the west bank of the Wissahickon and straddling much of the present-day driveway to "New" Erdenheim, this larger parcel was at some point between 1744 and 1761 transferred to William Barge, Mary Barge, Joseph Wollen, and Debora Wollen, before being sold on March 17, 1761, to William Dewees, a prominent landowner in Whitemarsh and Springfield Townships during the eighteenth century. Nine years later, though, Dewees must have suffered some sort of financial reversal, as the land was seized by the sheriff for the City and County of Philadelphia and conveyed to Anthony Williams Sr. on March 27, 1770. Fifteen years later, on April 9, 1785, Anthony Williams and his wife conveyed this same parcel to their son, Isaac Williams.[36]

Isaac Williams was not yet done acquiring land, however. Less than a year later, on February 2, 1786, Isaac Williams purchased 150.5 acres of land from the estate of Peter Knorr for 1,000 pounds ("it being the best price that could be gotten"), who had obtained this parcel in 1762 from Alexander and Abigail Colley. This lot included the land on both sides of what would become lower Flourtown Road (running northeast from its intersection with present-day Thomas Road to its intersection with Stenton Avenue), as far southeast as the previously described Anthony Williams parcel and almost as far northwest as present-day Joshua Road.[37]

Shortly thereafter, on July 16, 1793, Isaac Williams also acquired two adjacent parcels fronting on the north side of upper Flourtown Road for 300 pounds. Both were part of the estate of Jonathan Robeson, having been conveyed to him on February 19, 1776, by William and Rachel Dewees. The first, larger, quadrilateral parcel consisted of 25 acres, nine square perches, and bordered the old Peter Knorr parcel to the northeast, Christopher Robins's land to the northwest, Flourtown Road to the south, and the second triangular parcel in the transaction. This latter plot, comprising 7 acres, sixty square perches, also bordered the old Knorr parcel to the northeast and terminated in a point at the intersection of Flourtown Road and the old Knorr parcel (at the southeast-turning elbow just before Thomas Road).[38]

The deed records tell us that the first parcel contained a large "messuage or mansion house," which at the date of Isaac Williams's will (December 22, 1834) was Williams's primary residence. This mansion house, no doubt, is the large I-house positioned on the north side of Flourtown Road today, about halfway between Joshua Road and the intersection with Thomas Road. Given that the deed for the Robeson to Williams transaction on this parcel in 1793 references only the land and no house or "messuage," it seems reasonable to conclude that the surviving home was built by Williams shortly after the purchase.[39]

How Isaac Williams came to possess the remainder of the old Anthony Williams (the elder) lands is less clear. Some sources inaccurately suggest that Anthony Williams died sometime before 1780, when his son Isaac was taxed on 470 acres of land in Whitemarsh, a portion of which he lived on as a tenant, but the genealogical record is clear that Anthony Williams Sr. died in Bristol

Township on August 11, 1793, at the age of eighty-two, at which time he apparently conveyed the remainder of his existing lands in Whitemarsh to his son Isaac.[40] In the years after 1780, however, it becomes increasingly difficult to trace Isaac Williams's land ownership through the tax records, as land that he leased was often taxed to the leaseholder rather than to Williams. In 1785, for example, Isaac Williams is listed as a farmer with 230 acres and a dwelling, four horses, and six cows. By the following year, though, he is assessed for 375 acres and a dwelling, plus four horses and six cows—a figure that does not fully account for his purchase of the Knorr parcel that year. Meanwhile, the Anthony Williams parcels are listed separately at 170 acres and a dwelling—far less than what he actually possessed—plus two horses and three cows. In 1793 the figure for Isaac Williams was 346 acres plus two dwellings, one of them almost certainly the Isaac Williams House on Flourtown Road, which Isaac Williams may have built that year. By 1798 and 1803, however, Williams's holdings were down to 280 acres and a single dwelling, perhaps reflecting his lease or the sale of some other portion of his property, including a dwelling, while Joseph Williams—the nephew of Isaac Williams and son of his brother Anthony Jr.—was taxed on 170 acres and a dwelling, three horses, and six cows—precisely the same size of the Anthony Williams parcel listed in 1786, the livestock excepted.[41]

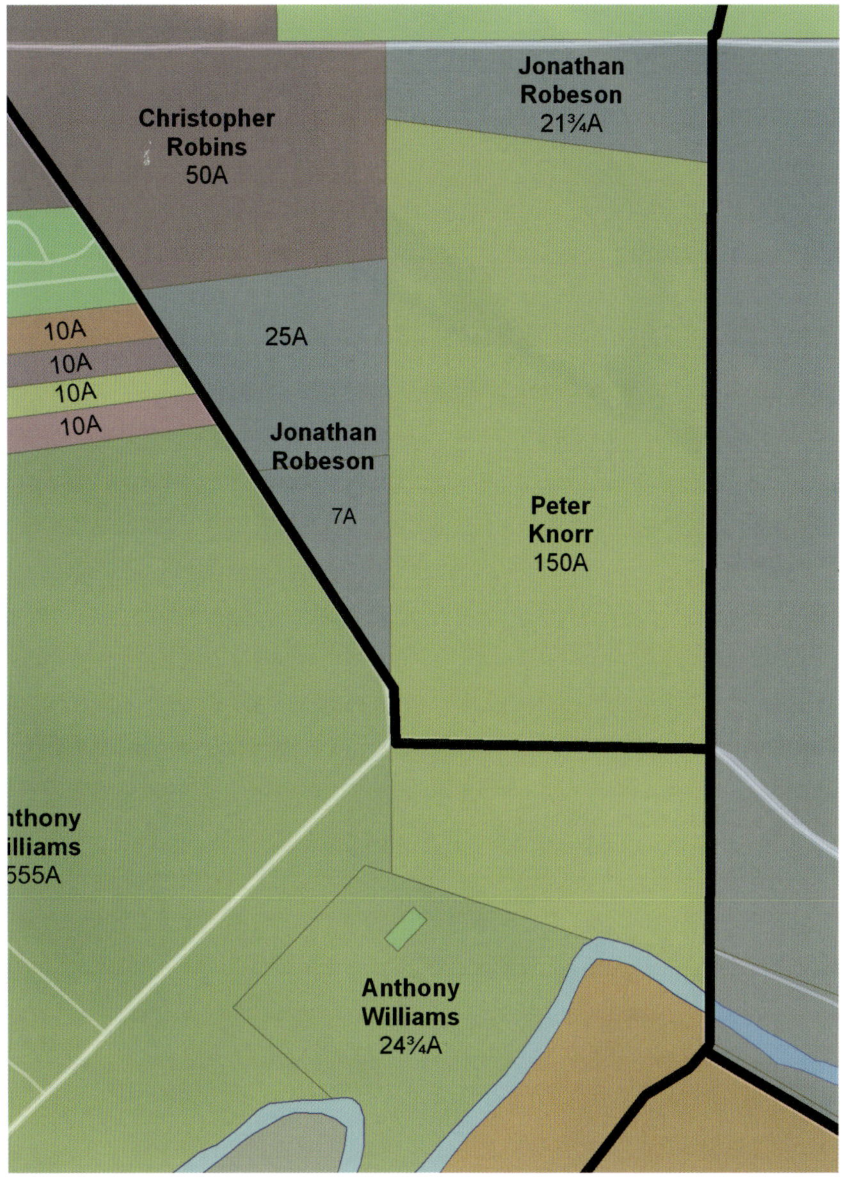

Detail from J. M. Duffin, *Mapping West Philadelphia: Landowners in October 1777*, University of Pennsylvania Archives. Isaac Williams acquired the large Peter Knorr parcel at center, the smaller Anthony Williams parcel at lower right (only a portion is showing), and the two Jonathan Robeson parcels along Flourtown Road during the 1780s and 1790s.

The Remaining Erdenheim Lands

By 1800, the vast majority of the land that would one day be Erdenheim was under the control of three principal landowners: William Streper (or Streeper), George Hocker, and Isaac Williams. The small half-acre inholding, set aside several decades earlier for a burial ground, occupied a portion of one of Isaac Williams's parcels. Aside from these lands, however, there still remained four long but narrow strips of land at the northwest extremity of today's Erdenheim (Angus Tract), fronting Flourtown Road, as well as a portion of a fifth parcel, the original ownership of which has not been clearly identified. Starting from the west end of the present-day Erdenheim lands along upper Flourtown Road were the portion of the unidentified parcel, then three 10-acre parcels owned, in order, by John Fisher, Bernard Rapp, and John Andrew Bower. A slightly larger parcel (12 acres, twenty square perches) lay immediately to the east of Bower's and was owned by Frederick Miller, it having been purchased from William and Rachel Dewees on May 10, 1770, along with the water rights to a well on an adjacent 10-acre parcel. This parcel included a home and several buildings at the time of its conveyance.[42] Miller, who was listed in the 1786 tax rolls as a weaver, was a man of otherwise simple means, possessing only one horse and one cow that year, apparently having sold or butchered one of the two cows he had in 1785.[43]

FACING: Isaac Williams House, Flourtown Road, January 2012. Photo by Eric Plaag.

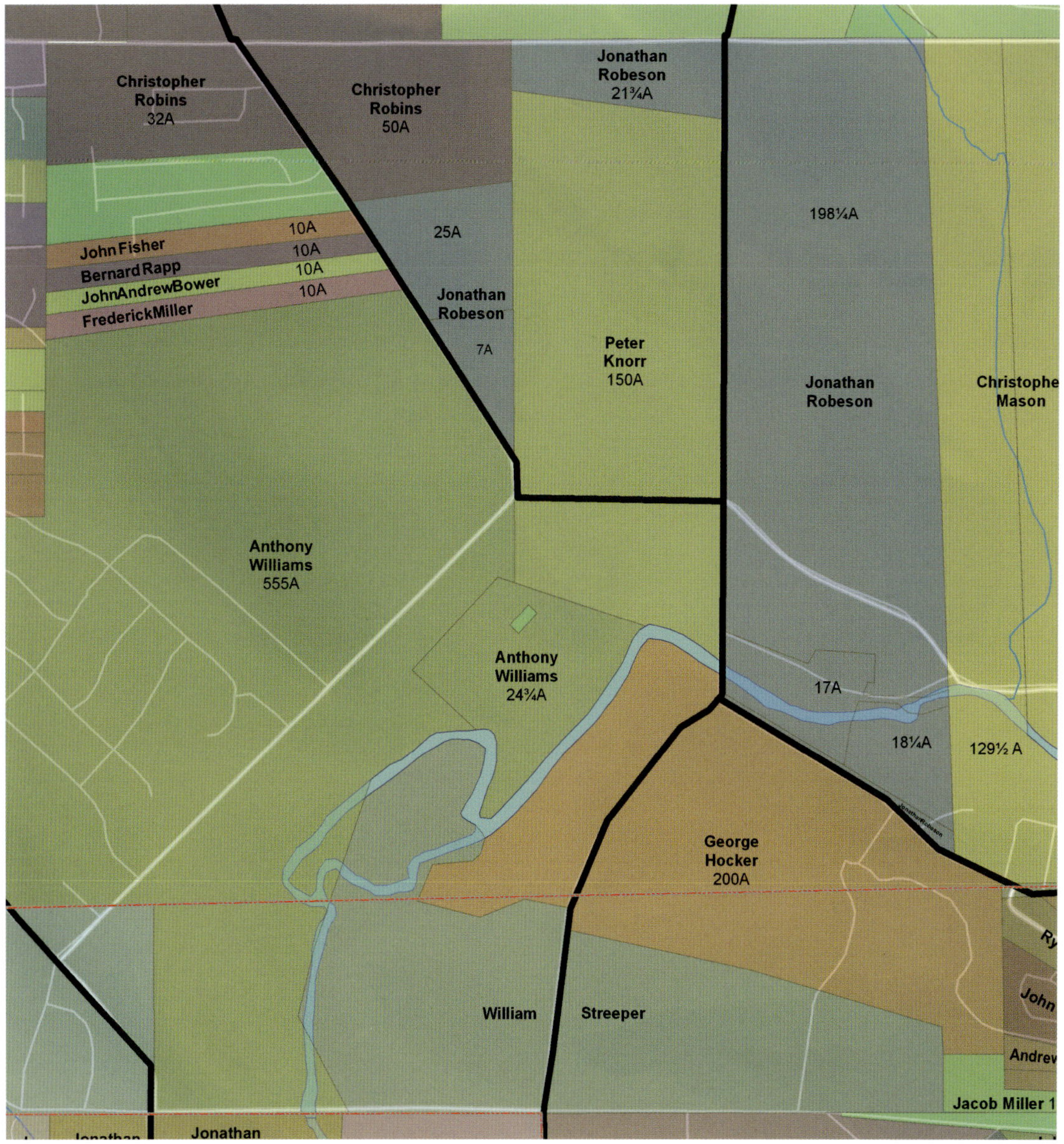

Overview of Erdenheim parcel ownership in 1777, from J. M. Duffin, *Mapping West Philadelphia: Landowners in October 1777*, University of Pennsylvania Archives.

In 1809 Miller in turn conveyed his parcel and messuage, along with the water rights on the adjacent parcel, to William Coulston Sr., who in turn purchased just over 3 acres of that adjacent 10-acre parcel in 1820. On August 4, 1838, William Coulston Sr. sold these lands as a single, 15.75-acre parcel to his son, William A. Coulston Jr., for $1,000—land that he would combine with other parcels he was purchasing in the vicinity at that time.[44]

Erdenheim and the American Revolution

Given Erdenheim's location on the outskirts of Philadelphia and in the midst of George Washington's activities during the late fall and early winter of 1777, it's somewhat understandable that historians of yore might be tempted to place notables from both sides of the American Revolution at Erdenheim. This is, in fact, what Lyman Horace Weeks did in describing the Erdenheim Stud in 1898: "Erdenheim has had a place in history quite apart from its connection with the breeding interests of the turf. It was one of the old colonial establishments of Pennsylvania, and in its solid gray stone mansion George Washington and other dignitaries at all periods in our country's history . . . were entertained."[45]

Sadly, there is only circumstantial evidence to suggest that Washington or his British combatants ever set foot on the land at Erdenheim, much less dined within the walls of the old house. One German map from the period, for example, suggests that American defensive positions in front of Whitemarsh (present-day Flourtown) may have extended to the southwest as far as the current location of the Carson School and perhaps as far as the surviving one-mile racing oval, with a battery set up near Robeson's/Cleaver's Mill (between West Mill Road and West Valley Green Road), but all the actual action took place along the Bethlehem Pike and east of there, well to the northeast of Erdenheim. By the time of Washington's retreat to Valley Forge, movement was almost entirely along the Germantown Pike, about a half a mile to the south and west of Erdenheim.[46] While it is possible that foraging soldiers from both sides tramped across Erdenheim's lands and may have stopped in to ask for food or supplies from Hocker, there is little or no evidence to prove it so.

Nevertheless old legends about Erdenheim's links to figures from the American Revolution persist. One of the most enduring is that Erdenheim played host to Gilbert du Motier, better known to Americans as the Marquis de Lafayette. Writing in the *Norristown Times Herald,* Edward Hocker, the noted regional historian, once tried to make sense of this legend. Acknowledging that Erdenheim's owners (presumably, at that time, the Wideners) "could tell nothing further as to the circumstances that brought Lafayette to Erdenheim," Hocker carefully analyzed the three occasions when Lafayette is known to have been in the immediate vicinity of Erdenheim. While he found each possibility remote, Hocker nevertheless concluded that the most reasonable possibility was that Lafayette briefly visited with General Henry Scheetz (about whom we shall hear more later) at Erdenheim during his 1825 visit to Chestnut Hill. "What would be more likely than that General Scheetz should arrange to greet General Lafayette at Erdenheim?" Hocker proposed. "Perhaps he did so arrange. And perhaps Lafayette did not come. And perhaps that gave rise to the tradition."[47]

The Williams School

The origins of this primary school are fairly clear, even though some nineteenth-century sources somewhat inexactly attribute its founding solely to Joseph Williams ("under the auspices of the Plymouth Preparative Meeting") in 1816. The full story is that on February 23, 1813, Isaac and Ann Williams, by then owners of much of the land on either side of lower Flourtown Road, sold to a group of men (including their relative Joseph Williams) for $1 ninety square perches of land for the "purpose of educating children in school learning." The deed was explicit about the use of this property, indicating that its transfer would be "for the purpose of erecting a schoolhouse thereon but in no way to be occupied as a Dwelling and also to be for the children to pass on any part of said lot," said property to be held in trust for the Preparative Meeting of Plymouth.[48]

About twenty years later, when the township adopted the common school system in 1836–37, the original building—an octagonal structure whose style was common among regional school buildings of the period and that was referred to alternatively as the Williams School or the "Eight-Square School"—fell under the watch of the township's school directors. William A. Yeakle, who served as the director of the school in 1866, personally supervised the demolition of the old octagonal building and the erection of a new structure for $1,900 that year, built by Isaac Brooks, a resident of Whitemarsh. The school property also had a new well reaching a depth of fifty-three feet.[49]

Early Roads at Erdenheim

Erdenheim is surrounded and bisected by some of the oldest roads in Springfield and Whitemarsh Townships. One of these early roads, Mill Road (sometimes West Mill Road), was first built to connect the village of Whitemarsh to the Bethlehem Pike (itself dating to 1698) in 1713. Stenton Avenue, known alternately as Welsh Road, Blue Bell Road, County (or Township) Line Road, and Lancasterville Road (named for the small community, sometimes called Wrangletown, that once lay just north of Erdenheim at the intersection of Militia Hill Road and Stenton Avenue), was first laid out in 1724. As was typical of this period, most of these roads had stone beds covered with dirt, and until 1804 their maintenance was the responsibility of local residents. Wissahickon Avenue, located at the southeastern edge of Erdenheim, was originally a toll road that was not constructed until 1854.[50]

Contrary to the current configuration, Flourtown Road—the old road leading from Norristown to Flourtown and known at times by either town's name according to which way one was traveling—did not originally follow its present course. Instead it passed south of the present intersection of Thomas and Flourtown Roads and just west of what would eventually be the Williams School lot, continued on beyond the old burial ground and across the present driveway

Eight Square School, Plymouth, from H. W. Kriebel, *A Brief History of Montgomery County,* 1923. The original Williams School probably looked something like this.

line of "New" Erdenheim, then forded the Wissahickon just below the present stone bridge of the private drive linking Thomas Road and Stenton Avenue (then known as the Lancasterville Road), then intersected Stenton Avenue at the lower driveway to the present Erdenheim property. When this road was later abandoned, probably in August 1793, a new road trace left the old road configuration near the Williams School, intersecting with the Lancasterville Road immediately north of the bridge over the Wissahickon, which was constructed by 1820 in response to an accident in which a Lancasterville lime burner lost a team of horses there trying to ford at high water. By 1853 this trace was abandoned by

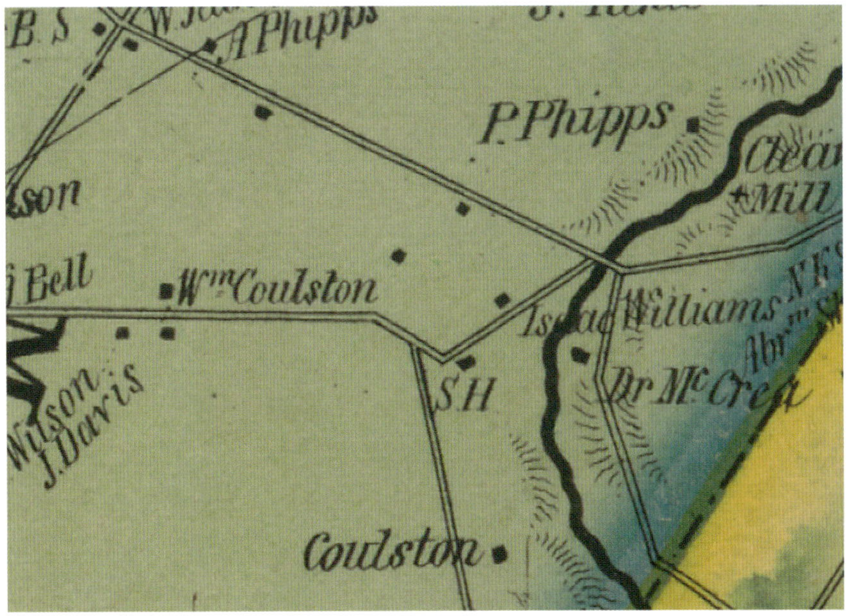

Detail from R. P. Smith, *Map of the Vicinity of Philadelphia from Actual Surveys,* 1852, Map Collection, Free Library of Philadelphia. The schoolhouse is denoted by "SH."

court order and a new one constructed following the present Flourtown Road trace across the Daniel Williams farm parcel (now known as the Sheep Tract).[51] Maps from the late eighteenth to mid-nineteenth centuries, meanwhile, make clear that Thomas Road (originally known as Wissahickon Road) was established prior to 1850 but not before 1777.

Two other old roads of note in the vicinity, according to tradition reported by Yeakle, served as shortcuts through then-existing parcels. The first of these branched off the first of the above-described Flourtown Road traces, passed just southeast of "New" Erdenheim, and continued southeast (running approximately parallel to present-day Thomas Road) to the Germantown Pike, ending just northwest of present-day Northwestern Avenue. The other ran diagonally to the south from the present limestone quarry on the west side of Stenton Avenue (at the northwestern extreme of Erdenheim's present boundaries) across the Daniel Williams farm (the present-day Sheep Tract) to a point just northwest of the present-day junction of Flourtown and Thomas Roads.[52] Quarrying, agricultural activity, and regrading of the land at present-day Whitemarsh Valley Country Club have obliterated nearly all traces of these old roads.

As the nineteenth century began, then, Erdenheim lay at the center of a network of roads linking Germantown, Norristown, and Flourtown, as well as dozens of smaller communities in the vicinity. At least two important mills were located on Erdenheim's periphery, subjecting the flow of the Wissahickon to both flooding and dry spells according to the whims of Erdenheim's neighbors as it meandered through Erdenheim's rolling fields. The parcels that would one day make up the Erdenheim that we know now were then the province of gentleman farmers, many of them with ties to the Quakers and a core sense of community. These were families subsisting off the land, for the most part, trying to make a living with their bare hands and extracting from the soil all that they believed

God was willing to give them. In the decades to follow, however, a new sort of owner would come to Erdenheim and its neighboring parcels, initiating changes that would one day bring Erdenheim and the valley of the Wissahickon to the attention of people all over the globe.

Notes

1. "Jasper Farmar, for 5000a. of Philadelphia," Copied Surveys, Series 17.114, Record Group 17, Records of the Land Office, book D65, page 293, Pennsylvania Historical and Museum Commission, Pennsylvania State Archives. See also "No. 845, Jasper Farmer, Jr., &c.," 1683, Philadelphia County Survey Book, letter F, volume 1, 356, Philadelphia City Archives. In many sources the Farmar family name is spelled "Farmer," sometimes interchangeably so within the same source. Some secondary sources report the date of this transaction as 1685, but both sets of survey documents clearly indicate 1683. *An important note regarding the British calendar pre-1752:* Prior to this year, the British Commonwealth and its possessions still used a modified form of the Julian calendar that began its legal year on March 25. As a result there were two calendrical problems at work, the first being the British legal adjustment, wherein dates were often indicated as being of a particular month of the year. Thus, the above noted survey indicates a warrant date of "ye 31st of ye 11th month 1683." Typically this date is rendered as "January 31, 1683 (O.S.)," where in the "O.S." designation indicates "Old Style." By the Gregorian calendar (which was more closely aligned with solar time), however, the year for this date would actually be 1684 in the "New Style." In addition the temporal discrepancy between the Julian and Gregorian calendars created a second calendrical problem, meaning that this date was actually February 10, 1684, on the Gregorian calendar. When the Calendar Act of 1750 finally went into effect, British subjects went to sleep on the night of September 2, 1752 (O.S.), and woke up in the morning of September 14, 1752 (N.S.). All dates prior to September 14, 1752, are rendered in this history in Old Style, while all dates after that date are rendered in New Style.
2. Buck, in Bean, *History,* 1138. Buck also reports that Whitemarsh was named for the "wide marsh [at] Farmar's Mill," located at the present site of Mather's Mill on the north side of the village of Whitemarsh. By 1720 the name Whitemarsh was being used interchangeably with "Farmarstown" for the township. See William A. Yeakle, "Whitemarsh," *Historical Sketches: A Collection of Papers Prepared for the Historical Society of Montgomery County, Pennsylvania* (Norristown, Pa.: Herald Printing and Binding Rooms, 1895), 20.
3. Yeakle, "Whitemarsh," 16–18. The original treaty is quoted in part here.
4. Yeakle, "Whitemarsh," 19–20. See also "Majr Jasper Farmer &c., war't for a devission of their lands," February 12, 1689 (O.S.), Copied Survey Books, D75, 10, Pennsylvania State Archives. A plat showing Catharine Farmer's parcel as well as the expanded lands of Edward Farmer (by now totaling 3,957 acres) can be found at "No. 820, Kath. Farmer, 1250 a., Edward Farmer," 1703, Philadelphia County Survey Book, letter F, volume 1, 348, Philadelphia City Archives. On Major Farmar's death, see Jean Barth Toll and Michael J. Schwager, eds., *Montgomery County: The Second Hundred Years,* volume 1 (Montgomery County Federation of Historical Societies, 1983), 793. As was common with many names in early deed records, variations of Catharine Farmer's name are rendered as Katharine, Katherine, and Catherine, sometimes within the same document.
5. Yeakle, "Whitemarsh," 23–24.
6. "The Governr, for 5,000 acres, Springfield, 1685," Copied Surveys, Series 17.114, Record Group 17, Records of the Land Office, book D85, page 116, Pennsylvania Historical and Museum Commission, Pennsylvania State Archives. For the Holme map, see Thomas Holme, *A Map of Ye Improved Part of Pennsilvania in America, Divided into Countyes, Tonwships and Lotts . . . ,* 1687, reprinted in appendix, *Pennsylvania Archives,* Third Series, Pennsylvania Historical and Museum Commission, Pennsylvania State Archives.
7. William J. Buck, "Springfield Township," in Bean, *History,* 1072.
8. See Nicholas Scull, *A Draught of Springfield Mannor in the County of Philadelphia,* 1732, Land Office Map Collection, Series 17.522, Map #3724, Pennsylvania State Archives. This map indicates that parcels were sold off beginning in 1689. A notation on this map indicates that a portion of Farmar's 247 acres was supposed to be exchanged with Nicholas Scull.
9. "Jonathan Robeson, 377 a., Resurvey'd, etc.," November 17, 1742, Copied Surveys, Series 17.114, Record Group 17, Records of the Land Office, book B7, page 206, Pennsylvania Historical and Museum Commission, Pennsylvania State Archives. Peter Robeson was a tanner and at one time served as Philadelphia's coroner. See Yeakle, "Whitemarsh," 61.
10. James Logan to Jonathan Robeson, "1st day of [--] mo, 1727/28," James Steel's Letterbook, 1715–1732, 83, Historical Society of Pennsylvania.
11. See Thomas Yorke to Peter Robeson, October 20, 1747, Deed Books volume H, book 2, page 307, Philadelphia City Archives. Yeakle, "Whitemarsh," 76, describes the 1895 land parcels that were encompassed by this vast transaction, which is helpful, given that the deed itself relies primarily on the customary deed descriptors of the time ("south forty seven degrees forty five minutes east one hundred and two perches and a half to a white oak sapling") rather than the more permanent landmarks often found in deeds today. There is no doubt, based on the deeds that will reference this transaction in later decades and centuries, that this transaction included a portion of the present-day Erdenheim lands. Nevertheless reconciling the 1742 Jonathan Robeson survey to this 1747 transaction has been deeply problematic, as none of the bounding owners listed on the 1742 Robeson survey align with those listed in the 1747 deed. This might be explained by the fact that Edward Farmar apparently bought and sold the same parcels on numerous occasions, as evinced by language in the Peter Robeson to Anthony Williams deed described below.
12. Yeakle, "Whitemarsh," 47. Yeakle credits the mill to either Peter or Jonathan Robeson. Jonathan Robeson reacquired the mill parcel as part of a 198.25-acre transaction in 1772,

according to the *Mapping West Philadelphia: Landowners in October 1777* project completed by J. M. Duffin and the University of Pennsylvania University Archives and Records Center (available at http://www.archives.upenn.edu/WestPhila1777/about.php), but his description as a "miller" in the deed for that transaction suggests that the mill was already standing. Nicholas Kline replaced the original mill building with a new one sometime in the early nineteenth century. Yeakle reported in 1895 that a steam engine was added to the mill during the previous year.

13. Peter Robinson [*sic*] to Nath. Sheppard, March 4, 1755, Deed Books volume H, book 6, page 434. Robeson is mistakenly listed as "Robinson" in the deed heading but correctly listed as "Robeson" elsewhere.
14. Peter Robeson to Anthy. Williams, March 4, 1755, Deed Books volume H, book 6, page 98, Philadelphia City Archives, and Yeakle, "Whitemarsh," 76. On Anthony Williams Sr., see Benjamin D. Shoemaker, compiler, *Genealogy of the Shoemaker Family of Cheltenham, Pennsylvania* (Philadelphia: J. B. Lippincott, 1903), 33–35, 57.
15. The quoted material is from Yeakle, "Whitemarsh," 29. An article addressing the title transfers is available at William R. Yeakle, "The Old Burial-Ground Near Erdenheim," *Bulletin of the Historical Society of Montgomery County,* 2:2 (April 1940): 95–104.
16. Yeakle, "Old Burial-Ground," 98–99.
17. Yeakle, "Whitemarsh," 29.
18. This positioning fits the deed descriptions for the parcel and matches the position given to the parcel in *Mapping West Philadelphia: Landowners in October 1777.*
19. Yeakle, "Old Burial-Ground," 102. As suggested by Yeakle, "Whitemarsh," 31, the larger stone probably belonged to John Nicholas Knight. Samuel Yeakle, in Samuel Yeakle, "'Erdenheim' and John George Hocker," in *Historical Sketches: A Collection of Papers Prepared for the Historical Society of Montgomery County, Pennsylvania,* volume 7 (Norristown: John Hartenstine, 1935), 114, remembered that the stone was engraved with the passage from Colossians 3:3–10, but the extreme length of this passage makes its entire inclusion in stone unlikely.
20. Yeakle, "Whitemarsh," 29–30. Yeakle makes clear that several unmarked graves were within the boundaries of the graveyard wall as well.
21. See Wm. Streper, et ux., to Geo Hocker, April 21, 1763, Deed volume 1, book 4, page 288, Philadelphia City Archives. The transaction for Nathan Sheppard to William Streper is described in this deed but does not list a volume, book, or page, and no such transaction could be found in the recorded deeds at Philadelphia City Archives.
22. Jean Barth Toll and Michael J. Schwager, eds., *Montgomery County: The Second Hundred Years,* volume 1 (Montgomery County Federation of Historical Societies, 1983), 613. This latter mill is one of the last surviving along the Wissahickon and presently makes up a part of the Morris Arboretum.
23. Streper to Hocker, April 21, 1763. Streper's name appears under variant spellings in land records of the day, including "Streyper," "Streeper," and "Streepers." As Yeakle reports in "Whitemarsh," 47–48, water levels along the creek were the subject of numerous lawsuits during the nineteenth century, as one mill's activity backed up water into another mill's operations. Hocker's first name, local historian Edward Hocker once speculated, was a baptismal name customarily dropped by German boys in later life. See Elizabeth Marshall Venable, *William Adam Hocker: Justice of the Supreme Court of Florida* (privately printed, 1941), 44–45.
24. "Where Robert N. Carson's Millions Will Rear a College for Orphan Girls," *Public Ledger,* October 24, 1907.
25. Yeakle, "Whitemarsh," 57–58, and Samuel Yeakle, "'Erdenheim' and John George Hocker," 108–9. The original deed to Hocker for this parcel indicates that the parcel included a "messuage" or dwelling house with surrounding buildings, a detail apparently included in the parcel transfer from Sheppard to Streper and mentioned in the deed transfer from Robeson to Sheppard. Local Historian Edward Mathews, writing in various local papers as "E.M.," concluded that whether or not Robeson built the house, it was certainly there by 1755. See E[dward] M[athews], "Erdenheim Stock Farm—Aristides Welch—Norman W. Kittson—Henry Hocker—Well-Known Property," undated (but c. 1910), uncredited clipping, Edward Mathews clippings scrapbooks, Historical Society of Montgomery County. For more on Edward Mathews, see Edward W. Hocker, "'E.M.'—Intensive Historian," *Historical Sketches: A Collection of Papers Prepared for the Historical Society of Montgomery County, Pennsylvania,* volume 6, Norristown Press, 1929, 234–41. Mathews's diaries are apparently on deposit at the Historical Society of Montgomery County as well.
26. *Johnson v. Hocker,* 1 U.S. 406 (Dall. 1789). Unfortunately the deeds for Hocker's transactions divesting himself of this property have not yet been located.
27. *Johnson v. Hocker,* 1 U.S. 408–9 (Dall. 1789).
28. John Streeper et al. to George Hawker, January 5, 1785, Deed Book 1, page 324, Historical Society of Montgomery County, and William Streeper, et ux., to George Hawker, August 3, 1784, Deed Book 1, page 322, Historical Society of Montgomery County.
29. Tax Dockets, George Hocker, 1786, 1793, and 1798, Montgomery County Archives. Unfortunately deeds confirming Hocker's land purchases in 1786 and land sale after 1793 have not been located. This is often a problem during this period; because Montgomery County was created out of Philadelphia County in 1784, many land transactions were not accurately recorded in the proper county for several years on either side of this date. In addition the indexing system for Philadelphia County's records went badly awry in the late eighteenth century, so that some transactions are indexed but can no longer be found using that index. Also, the amount of taxed land did not always equal the full holdings of a landowner, so it is difficult to draw too many conclusions from the acreage on which Hocker was taxed in Montgomery County after 1784. This was especially true when landholders leased land to tenants.
30. *Pennsylvania Gazette,* October 27, 1779, August 25, 1790, June 8, 1791, and September 28, 1791; see also *Pennsylvania Packet,* June 11, 1782.
31. William Adam Hocker, quoted in Venable, *William Adam Hocker,* 46. William Adam Hocker is also quoted as inaccurately describing George Hocker's above-described lawsuit as being a U.S. Supreme Court case. It was not.

32. Yeakle, "Whitemarsh," 58–59. Yeakle asserts that this story was relayed directly to him by Margaret Hocker Cress, George Hocker's daughter.
33. Yeakle, "Whitemarsh," 59. Sadly, the microfilm containing the criminal court sessions dockets for Montgomery County during this period has been stolen, and the Montgomery County Archives personnel would not permit direct access to the original dockets to confirm Margaret Cress's tale. William Adam Hocker relayed another version of this family story in Venable, *William Adam Hocker,* 46–47, placing the events of the robbery during the American Revolution and attributing them to one man rather than two.
34. Derived from an old Roman measurement term and used widely in Europe until the nineteenth century, a linear perch represented 16.5 feet or 5.5 yards, while a square perch was equivalent to 30.25 square yards or 0.00625 acres, thus yielding 160 square perches to an acre.
35. Register of Wills, Orphan's Court Docket, Johan George Hocker, OC 7778, Montgomery County Archives. The disposition of the remaining 8 acres, twenty perches of land has not been uncovered, but it may be the same land along the Wissahickon that was transferred to Isaac Potts and eventually to Lewis A. Lukens in the 1850s. This parcel appears to have been the result of an oxbow feature in the Wissahickon that created an island of sorts in the midst of the creek just above the old Piper's Mill millrace. As the creek straightened its principal course during the nineteenth century, it apparently created an outlier parcel on the west side of the Wissahickon.
36. Anthony Williams, et ux., to Isaac Williams, April 9, 1785, Deed Book 2, page 124, Montgomery County Archives.
37. Executors of Peter Kurner [*sic*] to Isaac Williams, February 2, 1786, Deed Book 2, page 196, Historical Society of Montgomery County.
38. Peter Robeson to Isaac Williams, July 16, 1793, Deed Book 7, page 759, Historical Society of Montgomery County. The lower part of Flourtown Road was not yet established at its present location.
39. Isaac W. Potts to William A. Coulston Jr., June 8, 1836, Deed Book 58, page 255, Montgomery County Archives. Williams is suddenly rated for a second house in the 1793 tax rolls, said dwelling described as being "added" that year. See Tax Dockets, Isaac Williams, 1793, Montgomery County Archives.
40. Yeakle, "Whitemarsh," 76, and Shoemaker, *Genealogy,* 33–35.
41. Tax Dockets, Isaac Williams, 1785, 1786, 1793, 1798, and 1803; Anthony Williams Estate, 1786; and Joseph Williams, 1803, Montgomery County Archives. On the Williams family, see Shoemaker, *Genealogy,* 57–58.
42. *Mapping West Philadelphia Landowners: October 1777* indicates that Fisher, Rapp, and Miller purchased their parcels in 1770, while Bower acquired his in 1772. Fisher and Bower both owned similarly shaped parcels nearby (all acquired in the 1750s) that fronted on the Germantown Pike to the west of Erdenheim. Rapp's parcel backed up to a much larger parcel fronting the Germantown Pike that he acquired in 1763.
43. Frederick Miller, Whitemarsh Township, Montgomery County Tax Schedules, 1785 and 1786, Montgomery County Archives. The dwelling Miller was taxed on in both years no longer stands.
44. William Coulston Sr. to William Coulston Jr., August 4, 1838, Deed Book 58, page 259, Montgomery County Archives.
45. Lyman Horace Weeks, ed., *The American Turf: An Historical Account of Racing in the United States* (New York: The Historical Company, 1898), 174.
46. See Johann Martin Will, *Plan des befestigten Lagers der Amerikaner beÿ Whitemarsh* (Augsburg, Germany: n. p., 1777), Library of Congress, Geography and Map Division. For a good overview of the Philadelphia campaign, see David G. Martin, *The Philadelphia Campaign: June 1777-July 1778* (Conshohocken, Pa.: Combined Books, 1993).
47. Norris [Edward Hocker], "A Lafayette Tradition about Erdenheim," *Norristown Times Herald,* undated clipping in Edward Hocker Scrapbooks, Historical Society of Montgomery County.
48. Yeakle, "Whitemarsh," 56–57, and Isaac Williams and wife to Several Appointed in Trust for the Preparative Meeting of Plymouth, February 23, 1813, Deed Book 29, page 167, Montgomery County Archives.
49. Yeakle, "Whitemarsh," 56–57, 62. Yeakle initially says the school was built "about the time of the Revolution," then corrects himself later in his narrative. Yeakle also lists several teachers from the old school that he can recall by name, including "Samuel Davis, Smith, Sappington, George A. Piper, Francis Murphy, a distinguished mathematician and an Irish refugee of 1798, [and] Joseph and Charles Willson." Earlier teachers included Thomas Paxon (listed as the school's first teacher), Goal Paxon, David Lukens, John H. Callender, Hughes Bell, and Jacob Paxon, according to Mahlon R. Ambler, whom Yeakle interviewed for his history.
50. Toll and Schwager, *Montgomery County,* volume 1, 615, 793.
51. Yeakle, "Whitemarsh," 55–56, 77. Yeakle recalls obliterating the oldest of these road traces on the grounds of the Williams School when he was director there in 1866 in order to smooth out the area for a larger playground. Portions of the second road trace, abandoned by 1853, are still quite visible in satellite imagery and align with present-day West Valley Green Road on the opposite side of Stenton Avenue.
52. Yeakle, "Whitemarsh," 56.

CHAPTER TWO

A Place Called "Ardenheim"

The sale of the majority of the old George Hocker lands to Casper Schlater, which occurred by auction in August 1822, was not completed by indenture until April 1, 1823, when Martin Hocker finally signed over the deed to 155 acres, 110 perches of George Hocker's Erdenheim parcel for the sum of $15,568.75 to Casper Schlater.[1] The disposition of the remaining 8 acres, twenty perches from Hocker's original estate is not clear from the archival record. In any case it's likely that Martin Hocker's sale of the bulk of Erdenheim's acreage to Casper Schlater may have been a convoluted but not at all uncommon ruse to pass the property without legal entanglements to Major General Henry Scheetz, George Hocker's son-in-law, for Schlater held on to the property for only eleven days before flipping the entire parcel to Scheetz on April 11, 1823, for the exact same price of $15,568.75.[2]

Schlater, who lived for many years in Upper Dublin Township, was born July 13, 1759, the son of Casper and Barbara Schlater, who came to Philadelphia from Rotterdam on the *Thomas Coatman* in 1752. At an unknown date he married Mary Seltzer, the daughter of Nicholas Seltzer.[3] Little is known about his early life, aside from the fact that he must have served in some capacity in the American Revolution, as he is listed as a "revolutionary soldier and companion in arms" to Daniel Miller in a published notice regarding Miller's service in the Revolution. Schlater vouched for Miller in 1829 along with Major General Henry Scheetz (about whom we will hear more shortly), both of whom were self-described as being "among the most respectable inhabitants of Montgomery County."[4] Schlater appears to have been deeply involved in the region's political scene and something of a rabid Democrat. A letter in the *Norristown Herald* on July 23, 1817, identifies him as a county commissioner and quotes him as saying that he "would rather vote for the devil, if nominated for an office by the Democratic Committee, than the best citizen in the State, nominated in any other way."[5] He also served as Montgomery County treasurer about this same time, and was "frequently called upon to act in the settlement of estates and the adjustment of disputes between neighbors."[6] His daughter Catharine married John Selser of Montgomery Township at an unknown date, although there were most certainly other children of Schlater's marriage.[7] Schlater died in 1835 and was buried at Boehm's Reformed Church in Blue Bell, where he had served as an elder, deacon, and trustee.[8]

FACING: Sheep grazing at the Sheep Tract. Photo by Eric Plaag, May 2012.

The Scheetzes

General Henry Scheetz (1761–1848) was the descendant of one of the oldest families in Whitemarsh, his father Henry (married to Catharine) having served for many years as a justice of the peace and as one of the associate judges of Montgomery County beginning in 1784, and his grandfather Henry having settled at nearby Sandy Run, where he operated several mills and engaged in the paper making trade. General Scheetz, however, took a different turn in life, earning his "rank" by serving as the commander of the Montgomery County brigade that turned out in 1798 to quell the Freas Rebellion in Lehigh County but arrived too late to make a difference. At an unknown date, Scheetz married George Hocker's daughter Elizabeth, who was eight years his junior, and the couple had nine children. Following the death of Elizabeth at the age of fifty-six in 1825, while the couple was likely living at Erdenheim, Scheetz married the widow of Peter Dager, who had been a frequent witness to land transactions and other financial matters in the township. One early description of Scheetz indicates that he was "tall and stoutly built, of dark complexion, but not black hair; dignified and manly in bearing, and wielding a marked influence in his party and on the general public till the time of his death."[9]

Scheetz owned several parcels of land throughout Whitemarsh at various points in his life, but around 1800, he owned a significant mill, which dated to the American Revolution, on the west side of the Wissahickon above Flourtown. One source described that building as having been "square to the creek, and was a long and low building, and presented . . . an antiquated appearance."[10] Scheetz later sold the mill to his son-in-law, Jonathan Wentz. Beyond his business pursuits at the mill, Scheetz served as a county commissioner in the late 1790s, a representative to the state assembly in the early 1800s, and in various other county appointments during the decades that followed, including as a director of the new poor house for the county in 1808 and again in 1811. He served as a presidential elector for Pennsylvania in the 1824 election, and in 1827 he failed in an election bid for the Pennsylvania Senate, losing by only forty-five votes.[11] At various times during the 1820s, he also held the post of director for the Bank of Montgomery County. He also saw service as a major general in the state militia during the War of 1812, commanding the Second Division, which was responsible for protecting the Dupont powder works at Wilmington, and apparently retired from the military as a brigadier general. He died in Valley Green, probably on land near his old mill (by then known as "Day's Mill"), on September 4, 1848, from pleurisy.[12]

Of General Scheetz's time at Erdenheim and his use of the farm we know almost nothing, aside from the fact that his wife, George Hocker's daughter Elizabeth, probably died there. For reasons that are unclear, General Scheetz sold the Erdenheim parcel to his son Jacob on April 1, 1830, for a substantial loss, securing only $14,000 for the sale of the 155-acre, 110-perch parcel.[13]

Of his son Jacob, who was born circa 1801, we also know comparatively little. At an unknown date, Jacob Scheetz married Christiana (sometimes "Christi Ann") Acuff, a woman his same age, but all records seem to suggest that they had no children of their own. Scheetz did follow to some degree in his father's military footsteps, serving as captain of the Second Troop of Montgomery County Cavalry sometime prior to 1839, and in 1830 he signed a published letter along with many of his neighbors, listing himself as a farmer of Whitemarsh and vouching for the reputation of Colonel John G. Watmough, whose name had been impugned by Daniel H. Miller, who alleged that Watmough had treated the employees of his farm in a cruel and unusual manner, "and was not allowing them either the usual rates of wages or the usual allowances of diet."[14] The 1850 Census—the first that lists household members and thus makes it possible to distinguish Erdenheim's Jacob Scheetz from others of the same name in the area—indicates that Scheetz and his wife were living by then in Northern Liberties, Ward 4, where Scheetz owned $35,000 in real estate; his occupation was listed as "gentleman." Other residents of his household included William Scheetz, a twenty-seven-year-old "merchant" who was probably Jacob Scheetz's nephew by his older brother Henry, as well as two "domestics" from Ireland.[15] Jacob Scheetz died on March 17, 1876, his last residence being at 1301 North Sixth Street in Philadelphia.[16]

As for Jacob Scheetz's life at Erdenheim, the record is rather thin. In 1833 he was taxed in Whitemarsh Township on 122 acres of land plus an additional 15 acres of "sport land," suggesting that Scheetz was either holding hunts at Erdenheim or engaged in some other public sporting pursuits using his land. This would account for only 137 of the 155 acres he is certain to have owned at Erdenheim, but such variances were fairly common in early tax records, often reflecting either leased lands or lands that were considered unusable. In addition to the acreage, Scheetz was also taxed on three dwellings, suggesting that the caretaker's/gardener's house on the west side of the intersection of West Mill Road and Stenton Avenue, as well as the superintendent's house on the west side of Stenton opposite the old stables, were standing by then. Scheetz was also taxed on six horses and nine cattle. In addition he paid taxes on two pleasure carriages.[17] In any case, by 1842, Jacob Scheetz was ready to part with the farm that had been his mother's childhood home. On October 5 of that year, Scheetz sold Erdenheim and its 155 acres, 110 perches to William W. Longstreth, a farmer "from New Jersey," for $23,500.[18]

William Longstreth Purchases Erdenheim

William Wilson Longstreth, born November 14, 1802, was the third son of Joseph and Margaret McKee Longstreth of Philadelphia, who ran a dry goods store there by 1796. Joseph Longstreth had been a Quaker, and his marrying out to a Presbyterian caused something of a scandal among the Friends. Nevertheless William Longstreth's father was permitted to remain a member, first of the Pikeland Monthly Meeting and later of the Philadelphia Monthly Meeting. He died just nine years later, however, from consumption at the age of thirty-three, when William was only four years old. In his absence Margaret Longstreth carried on the dry goods business in Philadelphia to support her family.[19]

Among William Longstreth's siblings was Morris Longstreth, who by 1837 owned a 240-acre farm at Valley Green, just north of Flourtown along the Wissahickon and very close to the old Henry Scheetz Mill. Whether this proximity to his brother's land was an inducement to William to purchase Erdenheim is not clear, but it may suggest that Jacob Scheetz and William Longstreth knew one another prior to the sale. William, however, does not appear to have had much interest in farming. As a child he mentored with Evans Rogers, a Philadelphia hardware merchant, and remained active in this trade in the city for much of his life, finally going into business for himself as a Market Street hardware seller under the firm name "Longstreth and Boldin"; there is good reason to believe that his business may have operated in the late 1860s as "William Longstreth and Company" (with Matthew Purcell) at 39 South Front Street, a firm that Longstreth's son William would run later on. If this is true, Longstreth's permanent residence was at 2017 Chestnut Street during this later period.[20] Earlier census

records suggest that he claimed residence in the North Ward of Philadelphia in 1840 and the South Ward in 1850.[21] Indeed, as became customary for many Philadelphia business owners during the nineteenth century, Longstreth may have regarded Erdenheim strictly as a summer residence, a place outside the city to which one might escape the city's various ills at the height of the season.[22]

Longstreth married three times. He wedded his first wife, Mary Bringhurst, on October 23, 1827, and the couple had four children: Joseph, James (who died in infancy), William, and Margaret, but Mary died just eight months after Margaret's birth, on September 26, 1835. Following her death Longstreth may have spent some time in New Jersey, as he is described as being a farmer from New Jersey in the Erdenheim deed and married his second wife, Margaret Collins Atlee, at Hanover, New Jersey, in February 1841. Atlee, the widow of Dr. Edwin P. Atlee, already had seven children from that union, all of whom were eighteen or younger, so the Longstreth family certainly could have used the space at Erdenheim when they either lived or summered there. Longstreth and Margaret added still another child to the horde shortly after their move to Erdenheim—Hannah, on July 12, 1843.[23]

Information on Longstreth and his activities in Whitemarsh Township is scant. He and his family did not live at Erdenheim during one of the federal censuses, so there is no way to confirm that his entire family lived there during that period, nor are there any surviving Philadelphia directories to counter this presumption. Whether or not Longstreth spent the bulk of his time in Philadelphia, he does appear to have been one of the secondary trustees listed on an 1847 deed for land set aside for a burial ground and meeting house located adjacent to the schoolhouse on Spring House Turnpike, near Fort Washington and close to his brother Morris's farm.[24] The Whitemarsh tax rolls indicate that he was taxed in 1843, the year after he purchased Erdenheim, on 120 acres (even though he possessed more than 155) "and improvements," four horses, thirteen cattle, and a pleasure carriage.[25] Eventually, though, either the farming life or the demands of maintaining a summer residence must have grown tiresome to Longstreth, for by 1849 he was described as a "gentleman" living in Philadelphia. On February 2, 1849, he sold Erdenheim for $18,371.12 to Dr. James A. McCrea; once again, the transaction was for the same parcel of 155 acres, 110 perches.[26] From the passport application he completed a week prior to the sale, apparently in preparation for a trip to Europe by steamer, we know that Longstreth was five feet ten inches tall, with a dark complexion, dark hair, and dark eyes.[27]

William W. Longstreth, from *The Longstreth Family Records.*

FACING: Superintendent's House, west side of Stenton Avenue. Jacob Scheetz was taxed on a dwelling at this location as early as 1833. Photo by Eric Plaag, January 2012.

A year later William and Margaret Longstreth were living in the Philadelphia South Ward with four of Margaret's children by Atlee, two of Longstreth's children by Mary Bringhurst, another female Atlee relative aged twenty-three, and two Irish servants.[28] Sometime after his sale of Erdenheim and the death of Margaret on April 3, 1861, at Germantown, Longstreth developed an interest in railroads and coal, and by 1863 he was serving as president of the Beaver Meadow Coal and Railroad Company. This line merged with the Lehigh Valley Railroad in 1864, and he remained president of both companies until he resigned for health reasons in 1868. Nevertheless he remained active with the board of directors of the rail lines and as chair of the finance committee. During this same period, Longstreth married his third wife, Ann W. Mather, the widow of Joseph Mather, on December 22, 1863. On June 3, 1879, just two days after the death of his third wife, Longstreth died at Philadelphia following a long illness. At the time of his death, Longstreth also owned a 110-acre farm at Hestonville, Pennsylvania, on the west side of the Schuylkill. He is buried with his mother at the Fairhill Friends Burying Ground in Philadelphia.[29]

Dr. James A. McCrea

Born in Philadelphia in September 1813, Dr. James Alexander McCrea came to his new farm in Whitemarsh Township already a middle-aged but very wealthy man. The grandson of an Irish immigrant and the son of a "wealthy India merchant," McCrea had studied medicine at the University of Pennsylvania and worked as a resident physician for two years each at the Pennsylvania Hospital and the Friends' Asylum for the Insane at Frankford before entering private practice and moving to a home given to him by his father at the northeast corner of Broad and Spruce Streets in Philadelphia. On November 5, 1839, he married Anna (or Ann) B. Foster. By all accounts, he feared succeeding in the medical profession, out of concern that he would burn himself out with "success." To take pressure off of his son, Dr. McCrea's father also purchased for him a signifcant interest in the Robesonia Furnace, an iron manufacturing operation in Berks County.[30] In spite of his father's assistance, Dr. McCrea's wealth and reputation were sufficient to earn him a place as a trustee for the Equitable Life Insurance, Annuity, and Trust Company in 1849.[31] Perhaps the most astonishing bit of evidence about Dr. McCrea and Erdenheim, though, is that the first recorded mention of the farm in Whitemarsh bearing a name of that sort suggests that the rest of us have been getting that name wrong for more than 150 years.

There are only scant sources to tell us about Dr. James A. McCrea's life at Erdenheim between 1849 and 1851. The 1850 Census reveals that by then their daughter Mary was nine and their son James two, and that a laborer from Germany and his wife lived at the farm, as well as a sixteen-year-old girl who probably functioned as a housekeeper or governess to the children.[32] Beyond that not much is known, but late in 1851, an article appeared in the *Pennsylvania Farm Journal* that suggested just how industrious Dr. McCrea had been with his agricultural pursuits at Erdenheim. Writing from "Whitemarsh, Montgomery County," on October 9, Dr. McCrea submitted a sample of the wheat he had grown that year on the farm, extolling the virtues of the seed he had obtained from a friend in the Wabash Valley, Cass County, Indiana. Crediting its origins to the Baltic, McCrea quoted his source to explain that the wheat crop yielded (using western farming methods) "thirty to thirty-five bushels per acre, . . . weighs sixty-four pounds to the bushel, and is considered the best in the country—commanding $1 per bushel for seed." After sowing the seed the previous October and harvesting it in July, Dr. McCrea then shared the results of his own tests with the grain, comparing it favorably with the Mediterranean variety he was accustomed to growing:

> I sowed the three bushels of seed on an acre and a half of ground which had been in potatoes, they having been sowed in the sod well manured with barn yard manure. The yield was fifty-two bushels. . . . It is a white beardless variety—bearing a full head not so long as the Mediterranean but equal in bulk upon a remarkably stiff straw, much the color of bright bamboo. Upon the 30th ultimo I seeded twelve acres with this wheat, having manured the oats stubble freely with barn yard manure, plowed three times, sowed broadcast two bushels to the acre and harrowed in.[33]

As if to underscore Dr. McCrea's capabilities in the fields, this same issue ran the list of exhibitions at that year's Pennsylvania State Fair; Dr. McCrea's two varieties of wheat were among the entrants.[34]

Dr. McCrea's accomplishments did not go unnoticed. The following year a reporter from the *Germantown Telegraph* paid a visit to his farm, which by then McCrea was calling "Ardenheim Farm." The farm was described as containing 156 acres of "fertile and productive land . . . not an acre of it that can not be cultivated," and Dr. McCrea's plantings were described as "the usual course of cropping throughout this region" and "met our unqualified approbation." In many ways, though, the reporter seemed more impressed with Dr. McCrea himself:

> The Doctor has nothing of the "fancy farmer" about him, and though he *did* come from Philadelphia, and *has* followed his profession, some five years of which being spent in the hospitals of the metropolis. He appears to take hold of everything at the right time, in the right place, and knows, as well as the best, what to do with it. His crops, as a general rule, are not merely the ordinary crops of the farmer, but they are, in many respects, *extraordinary* ones, as our readers may remember was the case with his wheat, last year, when over *forty bushels* to the acre were shown to have been raised by him.[35]

While the reporter acknowledged that the red weevil was likely to reduce the yield to twenty-five bushels that year, Dr. McCrea had taken precautions to prevent mischief in the granary. McCrea was also cultivating oats, grass, and barley that year, as well as timothy in a field "somewhat elevated above the other portions, amounting perhaps to three or four acres," in an area where drought damage had occurred the year before. McCrea also had 30 acres of corn—Dutton or eight-row variety, "the handsomest field of Corn we have seen this year." Most important, the reporter admired Dr. McCrea's planned, measured approach to cultivation: "Dr. McCrea, we are pleased to see, has adopted a *system,* which through the able assistance of his farmer, Mr. Neimann, he has been enabled to carry his plan of operations faithfully out."[36]

The reporter's observations also allow us to know a bit about the infrastructure at Ardenheim in 1852. His "barns and general out-buildings . . . were constructed with an eye to the uses of which they were to be applied. They are large, and are kept in the best possible condition." Dr. McCrea possessed little stock, keeping only enough to provide for his family. He had a shelter for the implements and wagons, a manure yard, and a good workbench. Dr. McCrea also reported that he turned a profit of 6 percent each year, and he estimated the value of the land at $200 per acre.[37]

While it might be tempting to explain away the *Germantown Telegraph*'s spelling of "Ardenheim" by imagining that Dr. McCrea possessed an Irish brogue that may have interfered with clarity, such a rationalization does not explain Dr.

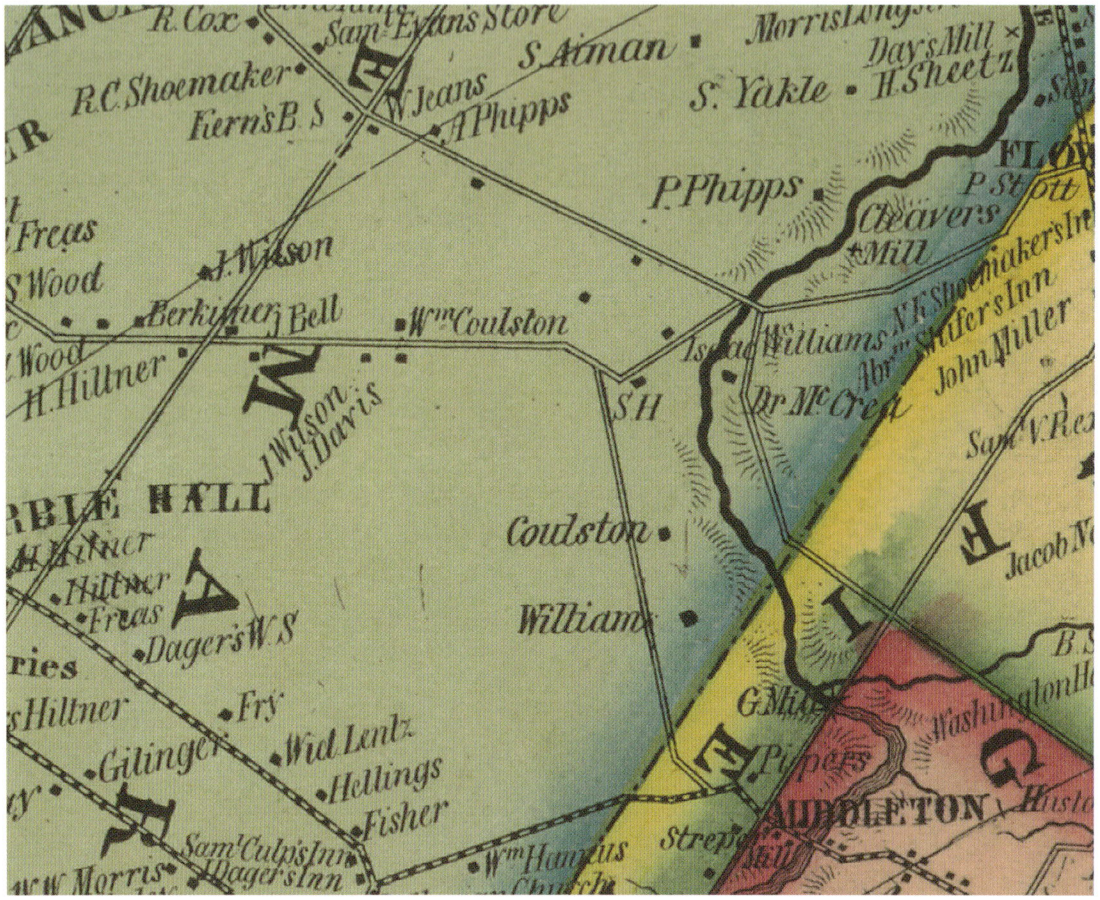

Detail from R. P. Smith, *Map of the Vicinity of Philadelphia from Actual Surveys,* 1852, Map Collection, Free Library of Philadelphia. "Old" Erdenheim—or "Ardenheim," as Dr. James A. McCrea called it—is indicated by the small rectangle just east of the Wissahickon and just to the left of "Dr McCrea."

McCrea's 1853 letter to the *Pennsylvania Farm Journal,* in which he signed his location of residence as "Ardenheim Farm, Whitemarsh tp., Mont. co., Pa." On this occasion Dr. McCrea was writing to discuss his method of corn cultivation as a means of competing for a premium announced by the Agricultural Society of the State of Pennsylvania. Using his methodology Dr. McCrea planted a field in two-thirds Dutton and one-third mixed variety of corn. Breaking up a sod of twenty years' growth in March, he planted in early May and did not employ manure. Unusual autumn weather undercut his efforts to demonstrate that the Dutton would mature earlier, and the resultant fodder was substantially higher than the usual average of nine feet, topping out at fifteen or sixteen feet high. The mixed variety proved to shell off about 4 per cent more by measure but 8 percent less by weight.[38]

Two months later Dr. McCrea was invited to take part in a public debate on corn planting, arguing for the relative merits of planting in hills or by the drill. While Dr. McCrea had planted in hills four feet by three and a half feet without manure, producing no stalk less than eleven feet in height, he nevertheless agreed that planting in drills and using poudrette, a quart to the hill, would have improved his crop.[39] Such appearances no doubt improved his reputation among his fellow farmers, as Ardenheim played host just three months later to a "trial of reaping and mowing machines," with two thousand spectators present, including one of the British Commissioners to the New York Exhibition.[40] In 1855 he also served as a judge in the "milch cow" category at the Pennsylvania State Fair.[41]

A more likely explanation for Dr. McCrea applying the name of "Ardenheim" to the farm was that he was making a nod to—or capitalizing on—the pop culture of the day. In 1848, a year prior to McCrea's purchase of the farm, George Lippard released his novel *Paul Ardenheim, the Monk of Wissahikon* to a mixed bag of acclaim and confusion. Lippard, a close friend of Edgar Allan Poe who manufactured the myth that the Liberty Bell was cracked the first time it was rung, was an immensely popular author perhaps known best for his novels that spun tales of secret societies, religious conspiracies, legends of the American Revolution, and morbid but titillating romance. His *Paul Ardenheim* stuck to this pattern, reimagining the legends and beliefs associated with the followers of Johannes Kelpius, a German apocalyptic mystic who forecast the end times for 1694 and brought his followers to the Wissahickon Valley in the belief that this was the best and safest place for them. When nothing apocalyptic occurred in 1694, Kelpius's followers continued to live and worship in caves along the Wissahickon. Lippard's *Paul Ardenheim* linked a remnant of Kelpius's followers with the events of the American Revolution.[42]

For whatever reason, however, by late 1855, Dr. McCrea had grown weary of farm life at Ardenheim. On December 31, Dr. McCrea and his wife sold Ardenheim and its 155 acres, 110 perches to Atherton Blight, a fellow Philadelphian, for $25,500, with the understanding that the remaining mortgage of $8,500 to William Longstreth would transfer to Blight.[43] From there Dr. McCrea remained active in a number of business pursuits in the Philadelphia area, including both the medical profession and farming. In 1859 he was appointed as a member of the Pennsylvania Board of Health and was named president of that organization for a five-year period beginning in 1863, and in 1876 he served as one of Pennsylvania's Centennial Commissioners. He also remained active with the Philadelphia Society for the Promotion of Agriculture and the State Agricultural Society, and he continued his ancillary business pursuits as well, serving as a manager for the Westmoreland Coal Company.[44]

But Dr. McCrea's closing years were not happy ones. While he remained active with the Board of Health and spent his summers with family at Rye Beach, New Hampshire, his own health did not hold for long. Beginning in the mid-1870s, he was overcome with what the papers described as a "disorder of the liver, which directly affected his brain, reducing him to a state of premature childishness, and tormenting him with that peculiar apprehension and overpowering dread of poverty to which so many business men have been subjected in their declining years." In the years to follow, he was beset by "chronic dyspepsia," "absence of mind," and insomnia. Finally, on June 22, 1880, he arose at his home at 2004 Delancey Place at his usual hour, prepared a bath, neatly folded and stacked his clothes "in the methodical way for which he was noted," locked the door, and slit his own throat with a straight razor.[45]

Atherton Blight

When Atherton Blight purchased Erdenheim from Dr. McCrea in December 1855, he probably saw it as the realization of a dream he had harbored for nearly two years. The son of Maria and George Blight, a wealthy Philadelphia merchant who owned substantial parcels of land along High Street and at a wharf along the Delaware River, Atherton Blight had lived a life of privilege. His father had died in 1834, just months before Atherton was born on November 16, and his father's will stipulated that as one of the five of the family's six children to survive through probate, Atherton's share of the estate would be $122,838.08.[46] We know almost nothing of his childhood, but twenty years later, in 1854, Atherton Blight was awarded his bachelor of arts in course from Harvard University.[47] While no surviving photograph of Blight has been found, we know from his preparations for a lengthy trip to Europe that year that he was five feet nine inches and had blue eyes and light brown hair.[48]

In October 1854 Blight departed with his friend and Harvard classmate Horace Howard Furness for Europe. After spending time in London and Paris, they soon joined their classmate James Savage Jr., and the three departed for Munich. The following year they continued their continental walkabout, doing their best to experience all facets of local culture and absorb the languages by complete immersion. That spring they saw Leipzig, Weimar, Berlin, Dresden, Switzerland, and ultimately Vienna.[49] While no personal account of his 1854 travels is known to survive, Blight's diary from the second part of 1855 and the first half of 1856 does. And it is this diary that reveals Blight's first dreams of owning a farm like Erdenheim. As Blight and his friends toured the large estates in Aigen, a district of Salzburg, he was so struck with their appearance that he remarked, "I have thought lately how charming it would be to have a small farm, green house, and pretty garden, devote my recreation hours to these, and my study hours to what relates to gardening and botany, and general literature."[50] Blight wrote frequently in his diary of his letters home to family, and it must have been in one of these that he mentioned his desire to purchase a farm. But Blight did not return to America to complete the transaction that occurred in late December 1855 or to witness the recording of the deed on Erdenheim in early 1856. Nor did he return home for the funeral of his sister, who died around this same time. Instead he remained on the Grand Tour, extending his visit back across Europe to Spain, then down into Egypt and Palestine, then up through Damascus and Beirut to Constantinople and into the Crimean Peninsula, where he viewed the battlefields of the recently concluded Crimean War. He then turned back west, stopping at Athens. His diary ends shortly after his brief stop on Corfu.[51]

It is not entirely clear when Blight did return to America, and it is curious that the farm purchase does not warrant a mention at any time within his diary.[52] We know that Atherton Blight purchased Erdenheim in December 1855—or at least that someone purchased the farm in his name and took possession of it in early 1856. We also know that Blight earned a master of arts degree, out of course, from Harvard, in 1858, and it appears that after that date he was working as an attorney primarily in Philadelphia and probably living there much of the time, too.[53] Indeed, as far as Erdenheim was concerned, Atherton appears to have grown weary of his dreams of farming life fairly quickly. Tax records indicate that both Atherton and his much older brother George were taxed in 1862 on property at the same residence on Township Line Road (one of the early names for Stenton Avenue), but George's share was much more substantial. Atherton was taxed on a one-horse carriage valued at $75, while George was taxed on 122 ounces of silver plate, a one-horse carriage, and two two-horse carriages, including one valued at $400.[54] The 1860 Census, meanwhile, lists Atherton Blight as an attorney living in Philadelphia with land valued at $200,000 and a personal estate of $50,000. He shared a house in the Ninth Ward of Philadelphia with his mother and two middle-aged servants from Ireland. The city directory for the following year shows him as an attorney with offices at 520 Walnut and a home at 1129 Chestnut, still more evidence that Erdenheim had become something of a lark for him.[55]

FACING: Cattle grazing at Erdenheim, with gardener's house visible in the distance. Jacob Scheetz was taxed on this building, located at the intersection of West Mill Road and Stenton Avenue, as early as 1833. In this image the Stenton Avenue bridge over the Wissahickon is visible at left. Photo by Eric Plaag, May 2012.

In the absence of hard, cold facts explaining Atherton Blight's purchase and use of Erdenheim, we must speculate as to what might have occurred. It may be that Blight's investment in the farm was in fact a cooperative venture between Atherton and his older brother George, rather than an effort by Atherton to become a farmer on his own. Long before Atherton's purchase of the farm, George Blight had been making a name for himself as a farmer throughout the state. As early as 1846, he was serving on the Committee of Arrangement for the agricultural exhibition of the Philadelphia Society for Promoting Agriculture, and he continued in that role in 1847.[56] In 1854, just prior to Atherton's purchase of the farm, George—who was listed as being from Germantown—won first prizes at the Pennsylvania State Agricultural Society Fair for his eggplants, sweet corn, squashes, and carrots, second prizes for his rye and parsnips, third prize for best amateur display of vegetables, and notice for his potatoes and corn.[57] Four years later, after Atherton had purchased Erdenheim, George Blight took home the top prize for this three-year-old Devon bull, Sir William, at the Chestnut Hill Agricultural Exhibition.[58] George's award-winning ways may also be the best explanation for why his name is often attached to Erdenheim on maps of the period, rather than Atherton's.

Or it may simply be that George Blight read his younger brother Atherton's sentiments about wanting a farm in one of Atherton's letters home, selected Erdenheim as a local, highly desirable farm, and purchased it in Atherton's name as a "welcome home" gift. We may never know for sure. It was at this time, too, that the caretaker's house directly opposite the intersection of West Mill Road and Stenton Avenue, as well as the superintendent's house directly opposite today's large stables on Stenton Avenue, began appearing on maps for the first time, although they may have been extant since Jacob Scheetz's ownership of Erdenheim.

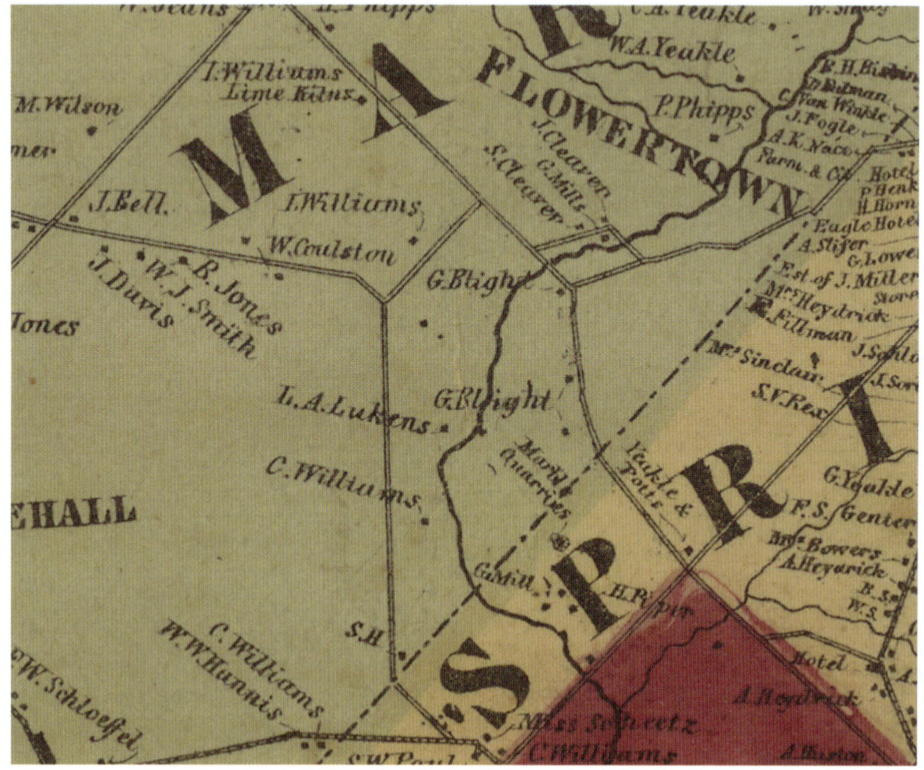

Detail from D. J. Lake and S. N. Beers, *Map of the Vicinity of Philadelphia,* 1860, Map Collection, Free Library of Philadelphia. Erdenheim is labeled with "G. Blight," rather than Atherton Blight's name. Note that four buildings are shown, including "Old" Erdenheim, the gardener's/caretaker's home opposite West Mill Road, and the superintendent's house opposite the present-day large stables on Stenton Avenue. The fourth building was likely a barn.

Regardless of his somewhat tenuous connection to the daily rhythms of Erdenheim, Atherton Blight's life in the years after he moved away from—or at least sold—Erdenheim was reflective of his privileged circumstances. In September 1862 Atherton Blight again left Philadelphia with his old friend Horace Furness, this time to answer General George B. McClellan's call for aides at the Union field hospitals during the American Civil War.[59] We cannot know for sure, but Blight may have adopted the same strategy as Furness, using the opportunity of travel on behalf of the Union to raise money for the Pennsylvania Sanitary Commission, on which he served. In any case Blight was back in Philadelphia by October 22, 1862, when he was witnessed signing the deed to Erdenheim over to Aristides Welch. Inexplicably, however, Blight still reported his address on his 1863 war draft registration as being at Township Line Road, at a property he had sold many months earlier.[60]

In 1872 Blight married Nina Greenough of Boston, Massachusetts, a woman thirteen years his junior and the daughter of famed American sculptor Richard Saltonstall Greenough.[61] The couple had three daughters that were by all accounts stunningly gorgeous, all of whom married well: the eldest daughter, Alice, wedded Sir Gerard Augustus Lowther, the British Ambassador to Constantinople, and thus became known internationally as "Lady Lowther"; the middle daughter, Edith, married William Payne Thompson, a wealthy Long Islander whose massive home "Westbury" dominated the Long Island social scene during the 1910s; and the youngest, Evelyn, tied the knot with Mahlon Sands, the son of a New York City drug importer for A. B. Sands and Company.[62] As for his own pursuits, Atherton Blight remained active with a number of philanthropic organizations in Philadelphia, including serving as a manager for both the Young Men's Home (formerly known as the "Newsboys' Aid Society") and the Pennsylvania Society for the Prevention of Cruelty to Animals, and on the Board of Council for the Church Home for Children.[63] His commitment to improving the lives of children also extended to involvement with the 1864 Great Central Fair for the U.S. Sanitary Commission, where he served on the Children's Committee and the Committee on Amusements during the Holding of the Fair.[64] In 1879 he also served as a director of the Pennsylvania Academy of the Fine Arts, and his marble bust—by father-in-law Richard Saltonstall Greenough—was on display as part of their annual exhibition.[65]

In 1880 the Blights moved their permanent home to a summer cottage on Bellevue Avenue in Newport, Rhode Island, where they became the center of the social scene during Newport's summer seasons, then returned to Philadelphia to spend their winters. In October 1897 Atherton Blight's wife, who had been ill with pulmonary tuberculosis for the better part of fifteen years, shot and killed herself at home shortly after a visit from her nurse. Atherton Blight thereafter spent the rest of his days in social and philanthropic pursuits, or with his children in Europe. He died suddenly from heart disease on November 4, 1909. His obituary described him as "in appearance, bearing, and manners a typical gentleman of the old school."[66]

The Williams Parcels

When Isaac Williams "the Elder," as he is often referred to in deeds of the period, passed away, the resultant flurry of land transfers amongst his various surviving relatives made a bit of a mess. Of Isaac, we do not know much, other than that he was born in Bristol Township on March 16, 1745, and married Phebe Meredith of Plymouth Township in November 1783. They had three children (Sarah, Elizabeth, and Isaac Jr.) before Phebe's death in January 1808. There is nothing to suggest that Isaac ever remarried.[67] His estate was massive, though, incorporating

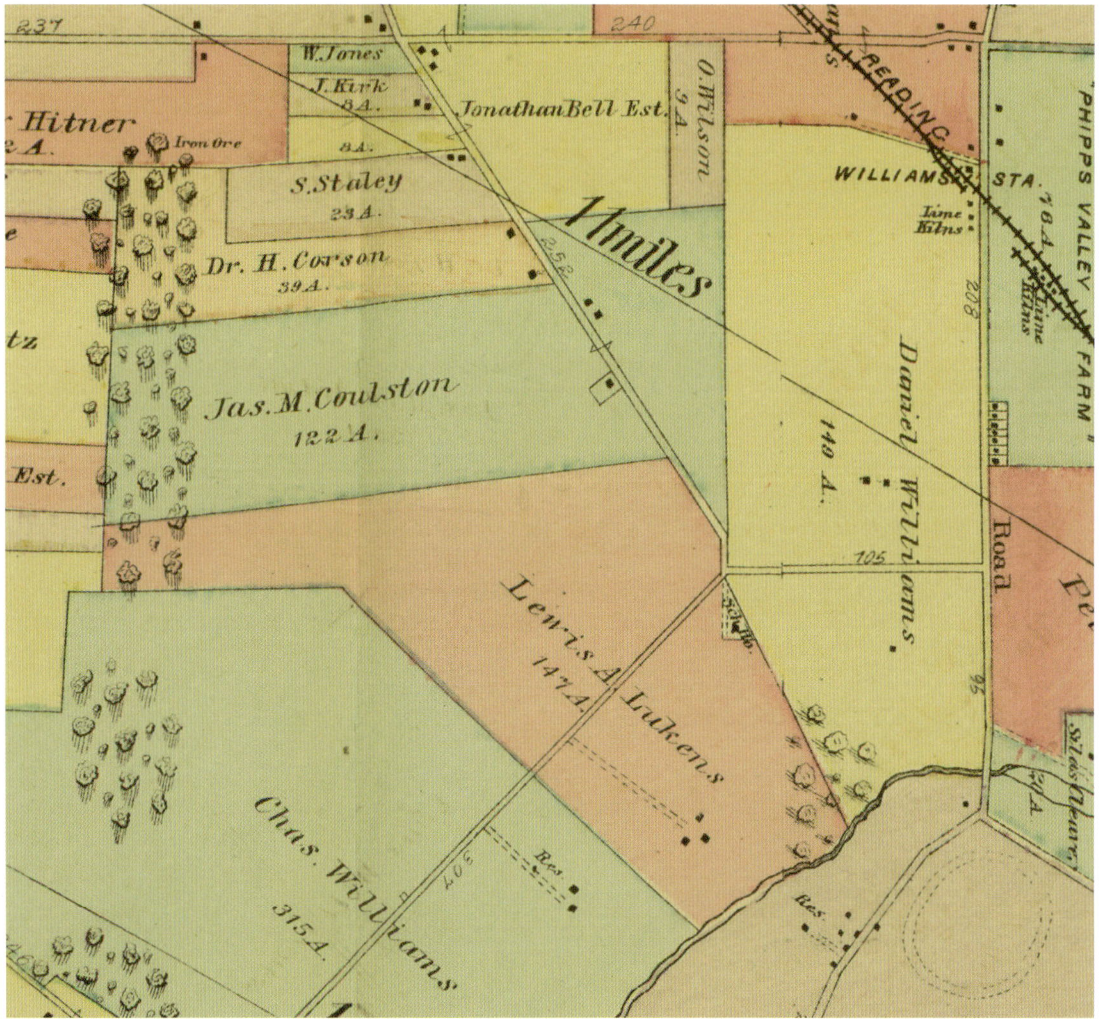

Division of the non-Hocker Erdenheim lands as of 1877, from G. M. Hopkins, *Atlas of Philadelphia and Environs,* 1877, plate 67, Map Collection, Free Library of Philadelphia.

much of the land on both sides of upper Flourtown Road, the vast majority of the land on both sides of lower Flourtown Road, and portions of the land on both sides of Thomas Road. The division of these lands among a host of children, grandchildren, and extended relatives resulted in the eventual creation of what really amounted to three separate "areas" of land, other than the original Erdenheim parcel, that today help make up Erdenheim Farm.

The first of these was the old Peter Knorr parcel, extending on both sides of lower Flourtown Road, now increased to 148 acres, 130 perches. By 1836 quarrying operations on the northeastern fringe of the Williams parcel, most likely developed at the direction of Isaac Williams, were sufficient to warrant the construction of a railroad station, Williams or Williams Station, along the Plymouth Branch Railroad, which ran from Cold Point to Springfield.[68] Isaac Williams, however, died before the train came through, passing on November 26, 1834, and this parcel remained tied up in the control of his various heirs for many years. These old Knorr/Williams lands then passed from the heirs of Isaac Williams to Anthony Williams III in April 1861, then again to Daniel Williams, son of Anthony Williams III, on December 14, 1863, in exchange for the payment of $5,500 to Anthony Williams's widow, Elizabeth Craft, and a series of other small financial commitments to his female siblings for the remainder of their lives.[69] Anthony Williams III was Isaac Williams's nephew by Anthony Williams Jr.; he married Elizabeth Craft at Abington Meeting in 1811 and died on February 15, 1868.[70] By all accounts, these lands had been intended as a gift to Daniel Williams from his father and were indeed bequeathed to him in his father's will dated 1863 but never formally transferred prior to his father's death. At some

FACING: Sheep Barn on former Daniel Williams parcel, north side of lower Flourtown Road, Sheep Tract. Photo by Eric Plaag, May 2012.

ERDENHEIM FARM
M

point prior to Daniel Williams taking possession of this parcel, the old Knorr tract was also expanded slightly southeastward to incorporate the old burial ground and an angled piece of land leading to the Wissahickon, although the deed and particulars for this transaction have not been found.

Of Daniel Williams, we know surprisingly little. He was born in Cheltenham Township on March 11, 1830, and he attended the Friends' School in Abington when not working on the family farm. Just prior to his twenty-first birthday, he married twenty-four-year-old Priscilla J. Kirk of Abington. His move to Whitemarsh did not occur until 1863, following his father Anthony's death, by which time Daniel and Priscilla had three children—John K., Alfred K., and Mary K. Once they were located at Erdenheim, they added two more—Edward C. and Howard H. The entire family claimed membership within the Society of Friends. Whether Daniel or one of the earlier owners of this parcel added the farmhouse that now occupies the north side of lower Flourtown Road is not clear, but Daniel Williams most certainly perfected the lime kiln operations that were quite active on the north end of his property on the west side of Stenton Avenue. In addition to his farming and lime operations, he had an interest in the iron works at Wilmington, Delaware.[71] By 1870 Daniel and Priscilla Williams owned real estate valued at $20,000 and personal property valued at $5,000. In addition to their five children, they shared their home with Irene Beecher, a twenty-two-year-old farm laborer, and Ann Spellman, a twenty-eight-year-old domestic from Ireland.[72]

As for the two Jonathan Robeson parcels along the north side of upper Flourtown Road that had been conveyed by Peter Robeson to Isaac Williams in 1793, and which contained the large "messuage or mansion house" that Williams built on this site before 1834, Williams left these parcels to his grandson Isaac W. Potts, along with another narrow plot of 30 acres of woodland located directly across Flourtown Road (on the south side) and extending southwest as far as the Barren Hill lots—all of it land that had once been part of the large Anthony Williams (the elder) tract during the seventeenth century. Potts appears to have been either a minor at the time of this transaction, or at least in severe financial distress, as all three plots were placed under a mortgage to be paid to his guardian, Anthony Williams (the younger), a move conditioned on the cutting and sale of all the timber on the 30-acre parcel. A year later, on June 8, 1836, Potts sold all three parcels to William Coulston Jr. for $3,750. Coupled with Coulston's purchase of the old Miller lands to the west from his father in 1838, Coulston controlled a total of nearly 33 acres on the north side of Flourtown Road and just under 46 acres on the south side. It was on this latter parcel, perhaps in 1839, that Coulston built the handsome residence located just west of the intersection of Thomas Road and Flourtown Road, on the south side of the latter.[73] At an unknown date thereafter, Coulston expanded his real estate on the south side of upper Flourtown Road, so that his total holdings by 1877 were 122 acres, thus composing the second "area" of Erdenheim land.[74]

William Coulston Jr. was born August 9, 1797, in Whitemarsh, and he married Ann Meredith (five years his junior) sometime prior to 1831. That year, on January 27, their son James M. Coulston was born. They had three children in all—James M., Elizabeth, and Hannah, but following Ann Coulston's death in March 1833, William was left to raise them on his own.[75] Little is known about the elder Coulston's use of the lands he owned, but by the 1860s, his son James was known to be mining portions of the land for its rich iron ore, as were his neighbors Charles Williams and Lewis A. Lukens, who by that time possessed

FACING: Daniel Williams House, north side of lower Flourtown Road, on Sheep Tract. Photo by Eric Plaag, January 2012.

James M. Coulston, from Bean's *History of Montgomery County*.

adjoining lands to the south that would eventually become part of Erdenheim.[76] James M. Coulston, meanwhile, married Tacy Amanda Freas on April 7, 1857, and their marriage yielded a total of twelve children (the last when their mother was forty-four years old), three of whom died in infancy. In 1863, following his father's death, James M. Coulston inherited a portion of his father's estate, then later purchased the portions belonging to his sisters. An active member of the Republican Party, Coulston also served as a school director in his district and as a director at the Montgomery Mutual Fire Insurance Company. He worshipped with the Plymouth Meeting Society of Friends and was a member of the Independent Order of Odd Fellows.[77] For a time, probably prior to her sale of the property to her brother, Hannah Coulston occupied the Coulston House on the south side of Flourtown Road, while James and his family lived in the Isaac Williams House on the opposite side of Flourtown Road.[78]

The third major "area" of land was the parcel directly opposite "Old" Erdenheim along the Wissahickon, on land that had originally belonged to the elder Anthony Williams during the eighteenth century. Tracing the chain of title on this parcel is especially difficult, but by 1849, Thomas Coulston, the brother of William Coulston Jr., and William Kettler shared the large tract at the present location of "New" Erdenheim, probably as tenants rather than owners. The 1850 Census lists Thomas Coulston, age thirty-five, and an older relative (perhaps his mother), Elizabeth Coulston, age seventy-two, as residing at the farm along with William and Mary Ketler, Adam Hoffman (age eighty-three), and four others who were probably domestics and farmhands. Coulston and Kettler were also known to have owned a farm in Gwynedd Township together, "and resided

FACING: Coulston House, south side of upper Flourtown Road. Photo by Eric Plaag, January 2012.

thereon the greater part of their lives."[79] The land adjacent to "Old" Erdenheim, however, most assuredly consisted of two tracts passed on by Isaac Williams to his daughter Elizabeth. The first of these comprised 147 acres, sixty-nine perches, while the second, located along the west bank of the Wissahickon and perhaps the missing parcel from the divestiture of the Hocker lands in 1823, consisted of 8 acres, sixty-seven perches. Whether the first of these parcels under Isaac Williams's ownership contained the house that would later be transformed into "New" Erdenheim is not known, but it was almost certainly standing by 1849, when Coulston and Kettler were residing there. When Elizabeth Williams Mather, who had married Joseph Mather in November 1810, passed away on May 23, 1844, these two parcels and the house were left to her various children. Over the next seven years, her children must have elected to lease out the land to Thomas Coulston and William Kettler, for they do not appear to have ever lived there.[80]

For whatever reason, Elizabeth Mather's various heirs finally elected to sell both parcels in 1851 to Lewis A. Lukens of Bridgeport, Upper Merion Township, for the sum of $16,311.75. [81] Lukens, who was born April 8, 1807, near Conshohocken in Plymouth Township, appears to have regarded the messuage and plantation at "New" Erdenheim as his permanent home only briefly. Both the 1860 and 1870 Census schedules list Lewis A. Lukens, an "Iron Man" and "Iron Master," respectively, and his family as residing at Conshohocken, where he possessed $43,000 in real estate and $1,200 in personal property in 1860 and relatively equivalent values in 1870.[82] At some point prior to 1877, Lukens also appears to have sold the smaller parcel to Charles Williams, his neighbor to the south, probably because the parcel represented a bizarre inholding within Charles Williams's land; the deed to this parcel contained an easement permitting the use of the private roadway on Williams's land out to Thomas Road. In any case maps of the period clearly show Lukens's holdings limited to 147 acres.[83]

Lukens was the descendant of Dutch ancestors who immigrated to Pennsylvania as early as 1683, and the Lukens name is ubiquitous in early documents of the colony. Lewis A. Lukens was the child of David and Mary Shepherd Lukens, and he initially received training to be a cabinetmaker. Abandoning that craft, he went into the lumber business with his brother Aaron for two years, then rented a forge in Lebanon County, where he produced malleable iron. About this time, on September 16, 1834, Lukens married Mary T. Wood, daughter of an ironworker from Conshohocken, and remained engaged in the iron business in Lebanon County for at least ten years. By 1849 they had seven children. After a short stint operating an iron store at Philadelphia (which he sold to his brother-in-law in 1845), Lukens returned to the lumber business at Bridgeport for several years. In 1851, following the purchase of the "New" Erdenheim property, he moved his family there, staying at the farm and engaging in agricultural enterprise until 1858. Farming life, as with so many other pursuits in his lifetime, must have grown tiresome, though, for Lukens abandoned it for the sheet iron manufacturing business that year with another brother-in-law, forming Alan Wood and Company, with locations at Conshohocken and 519 Arch Street in Philadelphia. He also served for many years as a director at the First National Bank of Conshohocken and as its president for four years. By 1877 he divested himself of these business interests and retired to Conshohocken.[84]

The last relevant parcel making up present-day Erdenheim lands was a 39-acre plot (only a portion of which is currently on Erdenheim property) located at the western or upper end of upper Flourtown Road and on its south side, just above James M. Coulston's land. It was owned in 1877 by Dr. Hiram J. Corson, the father of Susan Foulke Corson Lukens, who was the daughter-in-law of Lewis A. Lukens. Dr. Corson, who was born in nearby Hickorytown in Plymouth Township in 1804, maintained for most of his life a modest estate called

"Maple Hill" well to the north of the Erdenheim area, along Ridge Pike and Spring Mill Road. He appears to have been a member of the Plymouth Meeting and attended several of the Friends' schools and the University of Pennsylvania for his training. During the nineteenth century, he was probably the most prominent physician in Whitemarsh and published widely to great acclaim. His home at Maple Hill was also a well-publicized stop on the "Underground Railroad." He died on March 9, 1896.[85] It is most likely that Dr. Corson never used this parcel for anything other than farming, and he likely allowed his farmhands to live in the two modest homes on the property, one of which still stands as part of the Angus Tract of Erdenheim Farm.

Notes

1. Martin Hocker to Casper Schlater, April 1, 1823, Deed Book 39, page 66, Historical Society of Montgomery County.
2. Casper Schlater to Henry Scheetz, April 11, 1823, Deed Book 39, page 124, Historical Society of Montgomery County.
3. Ellwood Roberts, ed. "Henry Geiger Slingluff," *Biographical Annals of Montgomery County, Pennsylvania,* volume 1, New York: T. S. Benham, 1904, 143.
4. "Communication," *Philadelphia Inquirer,* August 27, 1829.
5. Quoted in Bean, *History of Montgomery County,* 506. For an example of his participation in political activities, see "Circular," *Franklin Gazette,* October 17, 1820.
6. Roberts, *Biographical,* volume 1, 143.
7. Bean, *History of Montgomery County,* 968.
8. Roberts, *Biographical,* volume 1, 143.
9. Yeakle, "Whitemarsh," 50–52; Samuel Wiley, ed., *Biographical and Portrait Cyclopedia of Montgomery County, Pennsylvania* (Philadelphia: Biographic Publishing, 1895), 39; M[oses] Auge, *Lives of the Eminent Dead and Biographical Notices of Prominent Living Citizens of Montgomery County, Pa.* (Philadelphia: by the author, 1879), 60–63; and Bean, *History,* 1142.
10. Yeakle, "Whitemarsh," 46–47.
11. "Extract of a Letter Dated Harrisburg, March 5," *National Gazette,* March 9, 1824, and "Misrepresentation," *United States Telegraph,* October 13, 1827.
12. Yeakle, "Whitemarsh," 46–52; Wiley, *Biographical,* 39; Auge, *Lives,* 60–63; and Bean, *History,* 1142. See *Map of the Vicinity of Philadelphia from Actual Surveys* (Philadelphia: R. P. Smith, 1852), Map Collection, Free Library of Philadelphia, for the location of this mill.
13. Genl Henry Scheetz to Jacob Scheetz, April 1, 1830, Deed Book 47, page 147, Historical Society of Montgomery County.
14. Bean, *History,* 192. There are numerous individuals named Jacob Scheetz who lived in eastern Pennsylvania during the mid-nineteenth century. For example, General Scheetz's son Jacob should not be confused with Dr. Jacob H. Scheetz of Pottstown, who was born in 1836. See Bean, *History,* 662. General Scheetz and Casper Schlater, curiously, had both vouched for Miller in the press in the fall of 1829. For an example, see Advertisement, *Philadelphia Inquirer,* September 18, 1829.
15. "Fourth Ward N. Liberties, Philadelphia, Pennsylvania," *1850 United States Federal Census,* roll 432-810, 194B, National Archives and Records Administration, Ancestry.com; on William Scheetz, see Auge, *Lives,* 61.
16. "Scheetz," *Philadelphia Inquirer,* March 23, 1876.
17. Jacob Scheetz, Whitemarsh Township, Montgomery County Tax Schedules, 1833, Montgomery County Archives.
18. Jacob Scheetz to William W. Longstreth, October 5, 1842, Deed Book 61, page 173, Historical Society of Montgomery County. Longstreth's second wife, Margaret, was from New Jersey, which may have led to the confusion associating Longstreth with that state.
19. Agnes Longstreth Taylor, *The Longstreth Family Records* (Philadelphia: Press of Ferris and Leach, 1909), 96–99.
20. Taylor, *Longstreth,* 197–99; and *Philadelphia, Pennsylvania, City Directory, 1867,* in U.S. City Directories, 1821–1989, Ancestry.com.
21. "Philadelphia North Ward, Philadelphia, Pennsylvania," *1840 United States Federal Census,* roll 482, page 206, National Archives and Records Administration, Ancestry.com; "South Ward Philadelphia, Pennsylvania," *1850 United States Federal Census,* roll M432-81, 121B, Ancestry.com.
22. Taylor, *Longstreth,* 198, says, "He lived in Philadelphia, but during the latter years of his life passed his summers at his country seat, a farm of about one hundred and fifty acres in the Whitemarsh Valley, some two miles beyond Chestnut Hill."
23. Taylor, *Longstreth,* 197–99.
24. Bean, *History,* 1145.
25. William Longstreth, Whitemarsh Township, Montgomery County Tax Schedules, 1843, Montgomery County Archives.
26. William W. Longstreth and wife to James A. McCrea, M.D., February 2, 1849, Deed Book 74, page 275, Historical Society of Montgomery County.
27. William W. Longstreth to James Buchanan, Secretary of State, January 25, 1849, *Passport Applications, 1795–1905,* NARA Series M1372, roll 24, National Archives and Records Administration, Ancestry.com.
28. "South Ward Philadelphia, Pennsylvania," *1850 United States Federal Census,* roll M432-81, 121B, Ancestry.com.

29. Taylor, *Longstreth,* 198; "An Old Merchant Dying," *North American,* June 3, 1879; "Mr. Longstreth's Death," *North American,* June 4, 1879; and Fairhill Friends Burying Ground Internment Records, *Historic Pennsylvania Church and Town Records,* reel 199, 250, Historical Society of Pennsylvania, Ancestry.com.
30. John H. Campbell, *History of the Friendly Sons of St. Patrick and of the Hibernian Society for the Relief of Emigrants from Ireland, March 17, 1771–March 17, 1892,* 1892 (reprint, Kessinger Publishing, 2006), 473; "Sad Events: Two Suicides in Philadelphia in One Day," *Philadelphia Inquirer,* June 23, 1880; and *Historic Pennsylvania Church and Town Records,* reel 649, Historical Society of Pennsylvania, Ancestry.com.
31. Advertisement, *Philadelphia Inquirer,* June 16, 1849.
32. Whitemarsh, Montgomery County, Pennsylvania, *1850 United States Federal Census,* roll M432-799, 142A, National Archives and Records Administration, available at Ancestry.com.
33. Jas. A. McCrea, M.D., "Communications," *Pennsylvania Farm Journal: Devoted to Horticulture, Agriculture, and Rural Economy,* November 1851, 233.
34. "List of Articles," *Pennsylvania Farm Journal: Devoted to Horticulture, Agriculture, and Rural Economy,* November 1851, 259.
35. "Ardenheim Farm," *Lewisburg Chronicle,* reprinted from the *Germantown Telegraph,* August 13, 1852, 1.
36. "Ardenheim Farm."
37. "Ardenheim Farm."
38. James A. McCrea, "James A. McCrea's Mode of Cultivation," *Pennsylvania Farm Journal: Devoted to Horticulture, Agriculture, and Rural Economy,* February 1853, 347.
39. "Corn Planting," *Pennsylvania Farm Journal,* May 1853, 58.
40. Untitled blurb, *Washington Reporter,* July 13, 1853.
41. "List of Judges at the State Fair," *Pennsylvania Farm Journal,* September 1855, 285.
42. For more on Lippard, see Mark A. Lause, *A Secret Society History of the Civil War* (University of Illinois Press, 2011), 22–23. For more on Kelpius, see Randall M. Miller and William Pencak, eds., *Pennsylvania: A History of the Commonwealth,* (University Park: Pennsylvania State University Press, 2002). On the Liberty Bell and Lippard, see Gregory M. Pfitzer, *Popular History and the Literary Marketplace, 1840–1920,* (Amherst: University of Massachusetts Press, 2008), 46. For reviews of *Paul Ardenheim,* see "Editors' Book Table," *Godey's Lady's Book,* January 1, 1849, 67; *Sartain's Union Magazine of Literature and Art,* April 1, 1849, 288; and "Reviewer's Department," *Nineteenth Century: A Quarterly Miscellany,* January 1, 1849, 157.
43. James A. McCrea and wife to Atherton Blight, December 31, 1855, Deed Book 100, page 313, Historical Society of Montgomery County.
44. John Hill Martin, *Martin's Bench and Bar of Philadelphia* (R. Welsh, 1883), 116; "Meeting of the State Agricultural Society," *Illustrated New Age,* March 23, 1866; "City Intelligence," *Philadelphia Inquirer,* January 7, 1864; Campbell, *History,* 473; and "Sad Events." McCrea's son, James McCrea, became president of the Pennsylvania Railroad Company in 1907, then served as president of multiple railroads simultaneously thereafter. He was also a trustee of the University of Pennsylvania. See "James McCrea," *National Cyclopedia of American Biography,* volume 4 (New York: James T. White, 1917), 446–47.
45. "Sad Events," and "Horrible Suicide," *Patriot,* June 23, 1880.
46. For details on Blight's birth, see "St. Stephen's Episcopal Church," *Historic Pennsylvania Church and Town Records,* reel 1079, Historical Society of Pennsylvania. On Blight's share of his father's estate, see George Blight, Transcript, 1834–1856, New York Historical Society. The catalog entry for this item is a bit misleading, as it consists of a transcript of Blight's original will, then a lengthy section on the 1856 disposition of the portion belonging to his daughter, Sarah Ann Rush.
47. "Commencement at Harvard University," *Boston Courier,* July 20, 1854. "In course" meant that a student completed the degree requirements without a break in studies. "Out of course" typically suggested a break in studies, most often between completion of the bachelor's degree and the start of master's work.
48. "Atherton Blight," October 4, 1854, *U.S. Passport Applications, 1795–1925,* Series M1372, roll 48, National Archives and Records Administration, Ancestry.com. For census details, see Philadelphia Ward 9, Philadelphia, Pennsylvania, *1860 United States Federal Census,* roll M653-1159, 24–25, Ancestry.com.
49. Horace Howard Furness Jayne, ed., *Letters of Horace Furness,* volume 1 (Cambridge, Mass.: Riverside Press, 1922), 27–49. This list of cities is a sampling. Blight's travels in the 1850s seem to have taken him nearly everywhere in continental Europe, the southern and eastern Mediterranean, and the Black Sea.
50. Atherton Blight, Account Book and Journal, 1849–1856, July 18, 1855, Historical Society of Pennsylvania. While portions of Blight's account book cover the period when he owned Erdenheim, no reference appears to the farm in his lists of expenses.
51. Blight, Account Book and Journal. His account of his travels is fascinating in its own right, both for his scrupulous accounting of the other Americans he encountered in his travels and his meticulous descriptions of his journey, much of it completed on horseback or on foot.
52. Blight's diary ends with the entry dated May 15, 1856. Blight's friend Furness returned to the United States in November 1856. Jayne, *Letters,* 102.
53. "Harvard University," *Massachusetts Spy,* July 28, 1858.
54. See "Atherton Blight" and "George Blight," *U.S. IRS Tax Assessment Lists, 1862–1918,* Pennsylvania, District 4, Monthly and Special Lists, Sept. 1862–April 1863, Series M787, roll 19, National Archives and Records Administration, Ancestry.com. Atherton's brother George was sixteen years older. See Harold Edgar Gillingham, *The Gillingham Family, Descendants of Yeamans Gillingham* (Philadelphia: Patterson and White, 1901), 39.
55. *Philadelphia, Pennsylvania, City Directory, 1861,* U.S. City Directories, 1821–1989, Ancestry.com; Philadelphia Ward 9, Philadelphia, Pennsylvania, *1860 United States Federal Census,* roll M653-1159, 25, National Archives and Records Administration, Ancestry.com.
56. "Agricultural Exhibition," *Farmer's Cabinet and American Herd Book,* July 15, 1846, 373; and "Agricultural Exhibition," *Farmer's Cabinet and American Herd Book,* August 14, 1847, 25.

57. "State Fair—Premiums Awarded," *Pennsylvania Farm Journal,* November 1854, 337–52.
58. "Local Intelligence," *Philadelphia Inquirer,* June 26, 1858.
59. Jayne, *Letters,* 111.
60. Jayne, *Letters,* 111; Atherton Blight to Aristides Welch, October 22, 1862, Deed Book 128, page 149, Historical Society of Montgomery County; Class 1, Fourth Congressional District, Pennsylvania, *Consolidated Lists of Civil War Draft Registrations, 1863–1865,* NM-65, entry 172, National Archives and Records Administration, Ancestry.com.
61. Town and City Clerks of Massachusetts, "Marriages Recorded in the City of Boston, 1872," *Massachusetts, Town Vital Collections, 1620–1988,* Ancestry.com.
62. "Atherton Blight," *New York Tribune,* November 5, 1909, 7; "W. P. Thompson, Widely Known Sportsman, Dies," *New York Tribune,* September 21, 1922; "Lady Lowther," *Washington Times,* May 20, 1909, 7; and "Mahlon Sands's Death," *New York Times,* May 9, 1888.
63. "Local Intelligence," *Philadelphia Inquirer,* February 8, 1862; "The Pennsylvania Society for the Prevention of Cruelty to Animals," *Evening Telegraph,* January 18, 1870, 8; "The Church Home," *Philadelphia Inquirer,* February 5, 1862.
64. Charles Janeway Stillé, *Memorial of the Great Central Fair for the U.S. Sanitary Commission* (U.S. Sanitary Commission, 1864), 160.
65. *Catalogue of the Fiftieth Annual Exhibition of the Pennsylvania Academy of the Fine Arts* (Philadelphia: n.p., 1879), 14.
66. "A Bullet Stilled Her Weary Heart," *Philadelphia Inquirer,* October 16, 1897; "Mrs. Atherton Blight," *New York Herald-Tribune,* October 13, 1897; "Atherton Blight," *New York Tribune,* November 5, 1909, 7; and "Atherton Blight Dead," *Philadelphia Inquirer,* November 5, 1909, 9.
67. Shoemaker, *Genealogy,* 37–38.
68. Toll and Schwager, *Montgomery County,* vol. 1, 794.
69. Ann Williams, et al., to Anthony Williams, April 2, 1861, Deed Book 123, page 147, Montgomery County Archives, and Will of Anthony Williams, proved March 11, 1868, Will Book 12, page 50, 06972/1309, Montgomery County Archives.
70. Shoemaker, *Genealogy,* 99–100.
71. "Alfred K. Williams," Roberts, *Biographical,* volume 2, 197–99.
72. "Whitemarsh Township, Montgomery County, Pennsylvania," *1870 United States Federal Census,* roll M593-1379, 443A, National Archives and Records Administration, Ancestry.com.
73. Isaac Potts to William Coulston Jr., June 8, 1836, Deed Book 58, page 255, Montgomery County Archives. Bean, *History,* 1159, indicates that the Coulstons were residing in this house by 1839.
74. G. M. Hopkins, *Atlas of Philadelphia and Environs,* 1877, plates 66 and 67.
75. Bean, *History,* 1159, and Roberts, *Biographical,* volume 1, 74.
76. Yeakle, "Whitemarsh," 77.
77. Bean, *History,* 1159; and Roberts, *Biographical,* volume 1, 75, 524.
78. See *Atlas of the County of Montgomery and the State of Pennsylvania,* Philadelphia: G. M. Hopkins, 1871, plate 66.
79. Whitemarsh Township, Montgomery County, Pennsylvania, *1850 United States Federal Census,* roll M432-799, 142A, National Archives and Records Administration, available at Ancestry.com, and Roberts, *Biographical,* volume 1, 74. This tract is labeled as "Thos. Coulston" on William E. Morris, *Map of Montgomery County, Pennsylvania* (Smith and Wistar, 1849).
80. Elizabeth Mather, et al., to Phebe W. Mather, et al., December 31, 1845, Deed Book 67, page 620, Montgomery County Archives. The smaller parcel, as described in a previous chapter, was probably the result of an island effect created by a shift in the principal course of the Wissahickon at some point early in the nineteenth century.
81. Phebe W. Mather and Nathan Harper and Wife to Lewis A. Lukens, March 11, 1851, Deed Book 79, page 222, Montgomery County Archives.
82. "Conshohocken, Montgomery County, Pennsylvania," *1860 United States Federal Census,* roll M653-1143, 165, National Archives and Records Administration, and "Conshohocken, Montgomery County, Pennsylvania," *1870 United States Federal Census,* roll M593-1378, 213A, National Archives and Records Administration, Ancestry.com.
83. See Hopkins, *Atlas of Philadelphia and Environs,* 1877, plate 67. The Lukens to Charles Williams deed could not be located.
84. Wiley, ed., *Biographical,* 234–35, and John W. Jordan, ed., "Jawood Lukens," *Encyclopedia of Pennsylvania Biography,* volume 1 (New York: Lewis Historical Publishing, 1914), 168.
85. On Corson, see "Jawood Lukens," 170–73; Bean, *History,* 352, 643–45.

CHAPTER THREE

Aristides Welch and the Erdenheim Stud

One of the most famous of Erdenheim's many owners, Aristides Welch, who purchased Erdenheim in 1862, is also one of the most mysterious figures attached to the property. Reclusive in the later years of his life, Welch frequently shunned the spotlight, often going by the name "A. Welch" in public matters, in spite of his widespread fame in the racing world. By all accounts, he refused to participate in the wave of biographical encyclopedia and *Who's Who*–style forms of self-promotion that proliferated during the late nineteenth century—both at the regional and the national level—and no publicly distributed photograph or engraving of him is known to survive.

Born September 28, 1811, Welch was the second child of Ebenezer Welch and Mary Heap Welch of Shippensburg, Pennsylvania.[1] Little is known about his early life, other than that he had two brothers (one of whom died in childhood) and two sisters. Welch was granted a purser's commission in the U.S. Navy on June 27, 1846, a move that came rather late in life, given that he was already

Aristides and Rachel Henrietta Jackson Armstrong Welch, probably circa 1856. Private collection of Mary Jane Echenhofer, great-granddaughter of Robert A. Welch.

FACING: Erdenheim's racing oval, with barber pole interval markers. Photo by Eric Plaag, September 2013.

thirty-five. About this time, on December 10, 1846, he may have married Sidney Virginia Francis, but their marriage ended by 1856, either as a result of Francis's death or by divorce.[2] By the following year, 1847, he was serving as a purser on the nine-gun steamer *Princeton,* the U.S. Navy's first screw steam warship, which saw service in the Home Squadron during the Mexican War. In 1847 the *Princeton* completed a two-year cruise to the Mediterranean, although it is unclear whether Welch was still on the vessel at that time. In February 1856 he resigned his commission with the navy, likely finishing his service on board the steam frigate *Merrimac.*[3]

Just a few months later, in Washington, D.C., Welch married Rachel Henrietta Jackson Armstrong, the daughter of General Robert Armstrong of Nashville, who was at least seventeen years his junior.[4] A little over a year later, Welch and his new bride purchased Blenheim, the 895-acre estate of the late Andrew Stevenson, former Speaker of the U.S. House of Representatives. Located near Charlottesville, Virginia, and one of the oldest farms in the western part of the state, Blenheim went to the Welches for $21,000, but it is not clear how long they owned the property.[5] Evidence of the rest of his onshore life before 1862 is scant, although family tradition holds that he was somehow invested in the Union Pacific Railroad and generated much of his wealth in this fashion.[6] Welch's name did surface in one of the early cases before the California Supreme Court. Listed as a joint purchaser of land in San Francisco with Squire P. Dewey (a land speculator whose investments were the subject of several suits) on a September 1853 transaction, Welch apparently refused to join Dewey in his efforts in 1857 to recover the entirety of the lot they held in joint tenancy. The court determined that Welch's failure to participate in the action served as a bar to Dewey's recovery.[7] Welch's name surfaced again in 1861, this time in the California newspapers, at the height of war fever, when his name was linked to a lobbying scandal in which friends (including Welch) of Lincoln's secretary of war, Simon Cameron, were accused of securing "close contracts" for military supplies through which "the Government is certain to be swindled out of vast sums." At the time Welch was described as living in California, but the 1860 United States Federal Census shows the Welches, their eight-month-old son Robert, Susan Beane (a domestic who appears on the 1870 Census at Erdenheim), and two other boarders residing in the Northeast Ward of Lancaster, Pennsylvania. Welch, who was described as a "gentleman," had a personal estate valued at $100,000.[8]

Aristides Welch Purchases Erdenheim

How Erdenheim came to Welch's attention and why he chose to abandon his California interests for Philadelphia ones following the start of the war remains a mystery. What is known is that on October 22, 1862, Welch (listed as being from Philadelphia) purchased a 155-acre, 110-perch parcel from Atherton Blight for $28,000, it being precisely the same parcel that Blight purchased from Dr. James A. McCrea in December 1855.[9] Why Blight sold the property is not known, but by June 1860, the twenty-five-year-old Blight, working as an attorney, was reported as living permanently in the Ninth Ward of Philadelphia with his widowed mother, Mary Blight, and two middle-aged Irish servants. Most likely, by 1860 Blight's efforts at Erdenheim had become more hobby than occupation, if he indeed ever lived there at all.[10]

In the years surrounding his purchase of Erdenheim, Welch also bought several other parcels in the vicinity that were not attached to Erdenheim, including a lot of more than 12 acres in Springfield Township that he later sold to Richard C. McMourtrie, and another lot of just over 3 acres on Militia Hill.[11] But Welch also devoted time and money to expanding the boundaries of Erdenheim beginning almost immediately after his arrival. His first purchase in this regard involved a nearly 67-acre parcel to the southeast of Erdenheim at the approximate present-day location of the one-mile racing oval and the Carson School, carrying the property as far south as the northeast intersection of Wissahickon Avenue and Stenton Avenue. The other was a nearly 10-acre parcel probably located just to the east of this.[12]

In the years that followed his initial Erdenheim purchase, Welch didn't grow just his holdings. He also grew a family. Son Robert A. Welch was born at Washington, D.C., just before the family's arrival at Erdenheim, probably in 1860, but a second son, William A. Welch, was born sometime in 1861 or 1862, probably at either Lancaster or Philadelphia. A third son, James A. Welch, arrived soon after, on December 14, 1863. By 1870, however, Aristides Welch had far more people living on his land than his immediate family, including three domestic servants and a cook (three of the four from Ireland), four farm laborers (all from England, Ireland, or Saxony), two gardeners (one from Ireland), and three hostlers, including a very young John McCloskey of Ireland.[13] That same year brought tragedy to the household, however, as Welch's wife, Henrietta, a member of the Presbyterian Church of Chestnut Hill, was buried at Philadelphia on July 2. She was only thirty-four years old.[14]

It is difficult to know with any certainty how Aristides Welch acquired his fortune, but one might guess from the available evidence that war profiteering and land speculation in California may have had something to do with it. As we shall see shortly, by 1866 Welch's Erdenheim was drawing national attention, and

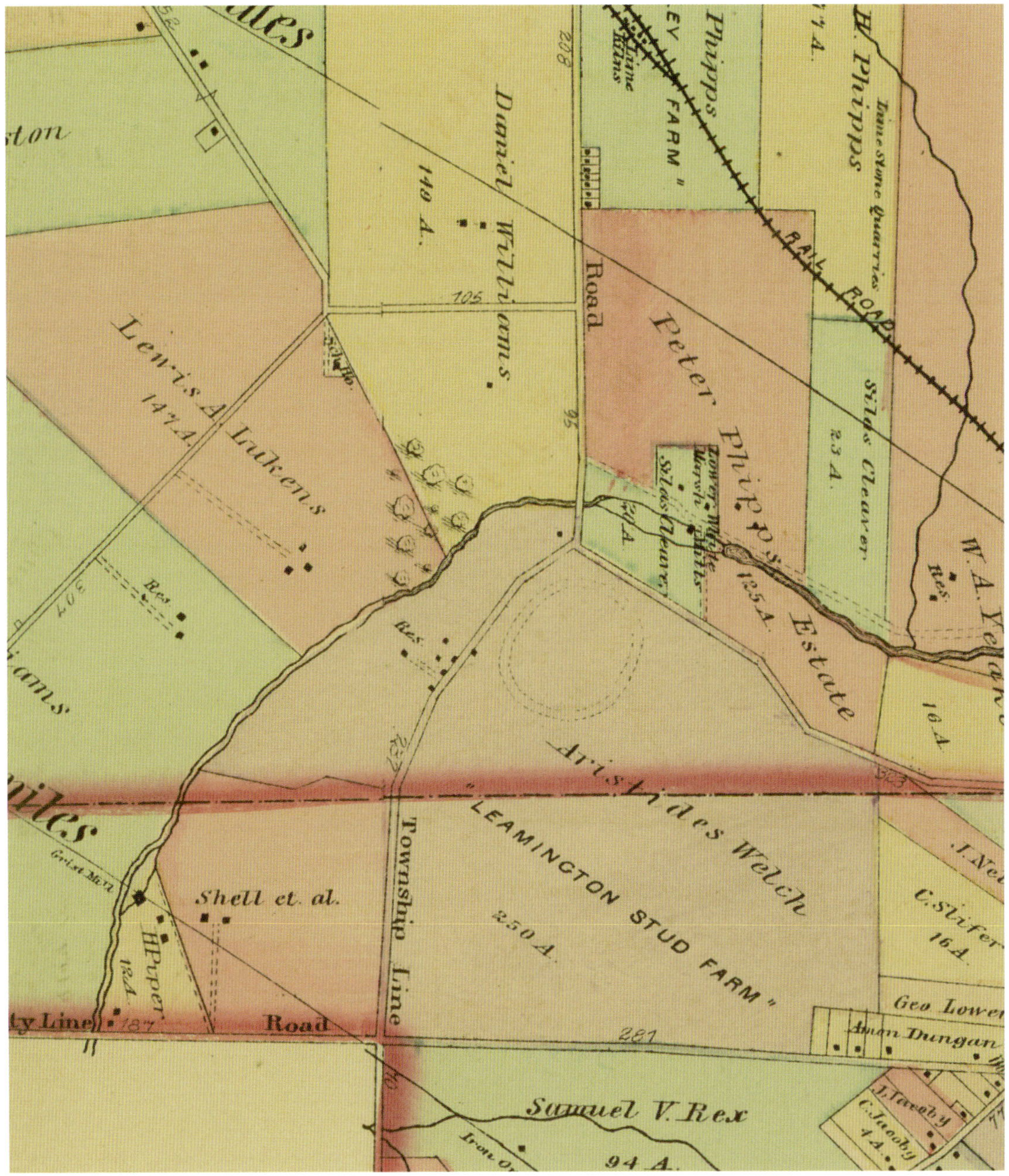

Aristides Welch's holdings at Erdenheim as of 1877. Detail from G. M. Hopkins, *Atlas of Philadelphia and Environs*, 1877, plates 66 and 67, Map Collection, Free Library of Philadelphia. Note the name "Leamington Stud Farm."

the national press described him as "a thorough gentleman, with an ample fortune."[15] While little survives in the way of documentation of Welch's connection with the great American statesmen of his day, local traditions insist that Welch frequently played host to such men even before he gained notoriety for his champion thoroughbreds. One of Welch's 1866 visitors may have been James Buchanan, who frequently visited Norristown in the years before he became president to discuss politics with a friend and business associate, John B. Sterigere; surely the fact that Welch was from Shippensburg, near Buchanan's birthplace, could not have hurt any friendship between them.[16] Welch's residence in the northeast part of Lancaster as of 1860, very close to Buchanan's home, Wheatland, may have also facilitated a friendship between them.

It may have been on one of these early trips to Norristown that Buchanan met Aristides Welch for the first time, perhaps even at Harry Connelly's, a private gentleman's club on the northwest corner of Seventh and Chestnut Streets in Philadelphia. As one nineteenth-century chronicler described it, Harry Connelly's was a place that "was an especial favorite with the Southerners," as well as "a rendezvous for people of diverse views and objects, a sort of neutral ground, where every body uttered his own ideas, and where every gentleman was tolerated." Such luminaries as Henry Clay, Daniel Webster, and John H. Crittenden were frequent visitors, and the aforesaid chronicler claimed to have met personally with James Buchanan and Aristides Welch, as well as other notables, "in this dark back room, with its low, cobwebbed ceiling . . . between 1845 and 1860." It

Aaron Wolff's old wine store, northwest corner of Chestnut and Seventh Streets, April 1859, photo by Frederick De Bourg Richards, courtesy of Library Company of Philadelphia. Harry Connelly's "back room" was located at the rear of this building. The building was demolished shortly after this image was taken.

was, the chronicler explained, a place where one could converse with "the kings of finance, of the stage, of the turf, and of politics. It was altogether a novelty, a spontaneous growth."[17] In any case, tradition holds, sometime during 1866, Buchanan spent some time at Erdenheim as Welch's guest, although no details of this visit survive in Buchanan's papers.[18] Other guests at Erdenheim during this period included August Belmont, Pierre Lorillard, and William Astor.[19]

Given the connections that Welch possessed by the 1860s, it should not be at all perplexing to understand how Welch also ended up at the center of one of the most fascinating national moments of the late 1860s—the bonding out of Jefferson Davis from prison following the American Civil War. In a political stalemate that mirrors some of the modern wrangling over Guantanamo detainees, the former president of the Confederacy spent eighteen months at Fortress Monroe following his 1865 capture, all without benefit of trial or parole, imprisoned under military custody that most Democrats and many Republicans believed was illegal. While Davis had numerous northern Democrats advocating on his behalf, one of his most surprising allies was Horace Greeley, editor of the Republican *New York Tribune*. Shortly after receiving a plea from Davis's wife, Varina, Greeley stepped up his efforts to either bail out Davis or encourage a trial.[20]

In the months that followed, Greeley assembled a rather bizarre cast of characters who agreed to stand as bondsmen for Davis. Among them was Gerrit Smith, a former financier for the abolition movement, the railroad robber baron Cornelius Vanderbilt, and twenty other men who together offered up the $100,000 bond, including Aristides Welch of Erdenheim. Having successfully secured a writ of habeas corpus on May 1, 1867, thus forcing the federal government to choose between trial and bail, Davis's attorneys arranged for him to be transferred to Richmond for the hearing. On the morning of May 13, the judge accepted the writ and indictment against Davis, then asked for the government to proceed with its case. Acknowledging that it was not prepared to prosecute Davis, the government allowed the defense's motion for bail. The judge encouraged each of the bondsmen, including Welch, to step forward and sign on Davis's behalf, then submit their share of the bond requiring Davis's appearance the following November. Davis and his wife then departed the courtroom, and several hours later, they left Richmond for Canada. Nearly two years later, in February 1869, the government formally dropped its case against Davis.[21]

So, why Aristides Welch? Piecing together that mystery is difficult. No evidence exists to demonstrate that Jefferson Davis and Welch knew one another or that they corresponded with one another, but Buchanan did maintain a close friendship and correspondence with Varina Davis over the years, despite various political difficulties between Buchanan and Jefferson Davis in the months leading up to the war. Varina Davis routinely pressed Greeley on raising bail for her husband as well, and it may be here that a bargain was struck, as Greeley considered "northern men" essential from both a practical and a political perspective. Jefferson Davis was also a close friend of John J. Crittenden, Welch's and Buchanan's old bar mate at Harry Connelly's. What remains most curious, though, is that Welch did not sign a second time for the renewal of Davis's bond in May 1868.[22]

Buchanan may not have been the only U.S. president to grace the halls of Erdenheim under Welch's tenure. Samuel Yeakle, speaking in 1922, reported that Welch and "Old" Erdenheim played host to Ulysses S. Grant for a three-day sojourn in 1868, presumably in the midst of Grant's election bid and allegedly to satiate Grant's abiding love for and curiosity about horses.[23] Other sources suggest that Grant was a regular visitor to Whitemarsh Township, on account of his mother, Hannah Simpson, having been born there and the presence of many relatives in Montgomery and Bucks Counties. Grant, some sources maintain, may have paid multiple visits to Welch at Erdenheim during this period.[24] Still other sources suggest that Grant participated in the auctions of yearlings at Erdenheim, particularly those of Welch's star stallion, Leamington, during the 1870s.[25] In any case Welch's connection to Grant may have been the impetus for a series of experiments conducted at Erdenheim sometime during the late nineteenth century "to combine the blood of wild horses of the western plains with that of an Arabian horse descended from one which the Sultan of Turkey had presented to General Grant." The intention was to create a superior breed of cavalry horses for the military, and the experiment apparently produced a "handsome, hardy breed," but it was abandoned when "they proved too expensive for cavalry service."[26]

The Erdenheim Stud and Flora Temple

The genesis and development of the Erdenheim Stud under Welch is not well documented, but perhaps the boldest move that Aristides Welch made in drawing interest in his endeavors was his purchase of the famed trotting mare Flora Temple in 1864 for the princely sum of $8,000. Flora Temple, it turns out, was intended as the linchpin to Welch's hallmark but controversial breeding strategy—the belief that "by using thoroughbred or nearly thoroughbred mares with trotting-bred sires, he could breed horses of greater speed and endurance than could then generally be found" among the trotters of the day.[27] By then Flora Temple's best racing days were behind her, but her career had been an astounding one.[28]

Her career did not get off to an auspicious start, however. Foaled in 1845 in Oneida County, New York, her breeder, a Mr. Tracy, hacked off her tail with his jack knife, cutting it so short that she earned the unattractive nickname "stump-tailed filly." Tracy found her "willful and unserviceable" and ultimately sold her for $13 when she was four years old. Her new owner, William H. Congdon, didn't think much of her either, passing her off to Kelly and Richardson for $68, thus managing at least to turn a modest profit on the deal. After several other successive owners dismissed her as a waste, she ended up in the care of George E. Perrin of New York, who purchased her for $350. Her first race in September 1850 was a victory, and by 1853, when she was owned by Hiram Woodruff, she was dominant on the track, winning seventeen of the twenty-one races she ran that year. Throughout the 1850s, and particularly under the subsequent ownership of William McDonald of Baltimore beginning in 1858, she continued her successful

FACING: N. Mitton, *The Celebrated American Trotting Mare, Flora Temple and Colt,* lithograph, circa 1870, Harness Racing Museum and Hall of Fame, Goshen, N.Y. This lithograph inexplicably attributes ownership of Flora Temple to "George Welch." The house in the background does not resemble "Old" Erdenheim, but given the presence of the stream in the central part of the image and Flora Temple on this side of that stream, the house in the background may be the Lukens house (i.e., "New' Erdenheim) on the opposite side of the Wissahickon. While there are some differences between this house's appearance and the known appearance of "New" Erdenheim circa 1900 after Carson's renovations, the general form and massing are in keeping with the house Lukens may have had at that time.

OVERLEAF: Louis Maurer, *The Celebrated Mare Flora Temple: "The Queen of the Turf,"* New York: Currier and Ives, 1860. Courtesy of American Antiquarian Society. This print commemorates Flora Temple's best performances and record-setting mile heat in 1859.

Nathaniel Currier, *Sontag and Flora Temple,* New York [?]: N. Currier, circa 1855. Courtesy of American Antiquarian Society.

Lith. & Pub. by Currier & Ives.

152 Nassau St. New York.

THE CELEBRATED MARE FLORA TEMPLE, "THE QUEEN OF THE TURF."

The Property of Wm. Mac Donald, Esq. Baltimore.

BEST PERFORMANCES.

DATE.	PLACE.	AMOUNT.	DISTANCE.	TIME.	BEATING.
1851 Sept. 9th	Union Course L.I.	Purse $30	Mile heats in harness	2:52 2:55 2:52 2:49	Whitehall & 3 others
1853 June 15th	Centreville Do	Match $2000	Mile "	2:29 2:37 2:32	Highland Maid
" June 29th	Centreville Do	Match $2000	Mile to wagons	2:26 2:32 2:32 2:35 2:36 2:35	Highland Maid
" July 15th	Union Course Do	Purse $1000	2 Mile in harness	4:59 5:01½	Tacony
" Aug. 20th	Utica N.Y.	Sweepstake $1500	Mile "	2:33½ 2:27 2:28½	Tacony
1855 Oct. 17th	Centreville L.I.	Match $2000	2 Mile "	4:53 4:57 5:21½	Hero (Pacer)
1856 Sept. 2d	Union Course Do	Match $1000	Mile "	2:24½ distancing	Tacony
1857 Sept. 2d	Elmira N.Y.	Purse $500	Mile "	2:26 2:27	Lancet & 2 others
" Sept. 5th	Elmira N.Y.	Purse $2000	Mile "	2:26½ 2:27 2:25	Lancet Do Do
1858 June 22d	Philadelphia Pa	Purse $1000	Mile "	2:31 2:27½ 2:29½	Lancet
" Nov. 4th	Adrian Mich	Match $1000	Mile "	2:30 2:28 2:28	Reindeer
1859 May 31st	Fashion Course L.I.	Purse $2000	Mile to wagons	2:25 2:27½ 2:27½	Ethan Allen

FLORA TEMPLE is a light bay mare 14 hands [illegible] inches high and weighs in trotting condition 835 lbs. was foaled in 1846 in the town of Sangerfield, Oneida County, N.Y. — out of "MADAM TEMPLE", by "ONE EYED HUNTER", he by the well known "KENTUCKY HUNTER". — She made her first appearance in public on the Red House track in 1850, beating the "WAITE PONY" in a single mile dash, — and in the same year on the UNION COURSE, L. I. under the name of "FLORA" won a race, mile heats in harness, beating "WHITEHALL" and three others. She is entered on the records of the turf in 95 RACES, winning 75, and receiving forfeit in 5, and 2 drawn.

BEST TIME: in harness mile heat 2:19¾ two mile heat 4:50½ to wagon, mile heat 2:25.

BEST PERFORMANCES.

DATE.	PLACE.	AMOUNT.	DISTANCE.	TIME.	BEATING.
1859 Aug. 9th	Eclipse Course, L.I.	Purse $1000	Mile heats in harness	2:23½ 2:22 2:23½	Princess
" Aug. 16th	Eclipse Course, L.I.	Purse $1000	2 Mile Do	4:50½ 5:05	Princess
" Aug. 23d	Boston Mass.	Purse $1000	Mile "	2:23 2:24½ 2:24	Princess
" Sept. 1st	Portland, Me	Purse $1000	Mile "	2:32 2:26½ 2:29	Princess
" Sept. 8th	Philadelphia, Pa	Purse $1500	Mile "	2:41½ 2:31 2:28	Princess
" Sept. 20th	Baltimore, Md	Purse $1000	Mile "	2:29 2:31 2:22	Princess
" Sept. 29th	Chicago, Ill.	Purse $1000	Mile "	2:31 2:31 2:26½	Princess
" Oct. 9th	Cincinnati Ohio	Purse $1000	Mile "	2:27½ 2:27 2:21½	Ike Cook
" " 15th	Kalamazoo, Mich.	Purse $2000	Mile "	2:23½ 2:22½ 2:19¾	Princess & Chance
" Nov. 21st	Union Course, L.I.	Purse $1000	Mile "	2:26 2:23 2:24 [illegible]	Geo. M. Patchen, Tobacco
" 24th	Union Course, L.I.	Purse $1500	Mile "	2:27 2:26½ 2:27 2:29½	Ethan Allen
" Dec. 1st	Baltimore, Md	Purse $1000	Mile "	2:27½ 2:26½ 2:25½	Ethan Allen

LITH. & PUB. BY N. CURRIER. 152 NASSAU ST. N.Y.

SONTAG AND FLORA TEMPLE.

AT THE HALF MILE POLE IN 1:13!!

IN THEIR GREAT MATCH FOR $ 2000. OVER THE UNION COURSE L.I. MAY 7TH 1855

MILE HEATS BEST 3 IN 5 TO WAGONS, WAGON AND DRIVER WEIGHING 300 lbs.

W. WHEELAN NAMED G. M. SONTAG .... 1. 1. 1.

W. PEABODY NAMED B. M. FLORA TEMPLE .... 2. 2. 2.

Time — 2:31 – 2:33 – 2:33

FLORA TEMPLE
FABRICA DE TABACOS
de la Vuelta Abajo
DE ANTONIO CABARGA
Calle del Empe drado N.47.
HABANA.
Prang & Mayer's Lith.

FLORA TEMPLE
FABRICA DE TABACOS
de la Vuelta Abajo
DE ANTONIO CABARGA
Calle del Empe drado N.47.
HABANA.
Prang & Mayer's Lith.

career, regularly improving on her record times and frequently outdueling the best trotting champions in the country. Her most impressive accomplishment, however, came on October 15, 1859, when her third heat on the mile track resulted in a shockingly fast, record time of two minutes nineteen seconds—a record that stood unsurpassed for the next eight years.[29]

Possessing lightning speed and establishing a record is one thing. Consistency is another, and Flora Temple was filled with it. Over the course of her eleven-year racing career (1850–61), the "Bob-Tail Nag" immortalized by Stephen Foster's minstrel song "Camptown Races" won ninety-five of the 112 races she participated in, finished second fourteen times, third once, and unplaced only twice. So fast and famous was Flora Temple that her name was synonymous with unthinkable speed; one clever newspaper editor, writing of the embarrassing Union retreat following First Bull Run in 1861, announced her imaginary defeat by the rapidly fleeing troops.[30] The dig was a curious one, given that the Union army, as a result of McDonald's southern sympathies, confiscated Flora Temple that year as contraband. When she was offered for sale in 1864 (after being returned to McDonald's estate), Aristides Welch purchased her for $8,000. To celebrate his purchase, Welch commissioned renowned artist Edward Troye to paint a portrait of Flora Temple and Kitty Temple, Flora Temple's daughter by Rysdyk, on July 22, 1868. The chromolithograph was later widely advertised for public purchase.[31]

By early 1866 it was clear that Welch was assembling one of the finest stables in America. That January a writer for *Turf, Field, and Farm* and Colonel David McDaniel of Richmond, Virginia, visited Erdenheim (which Welch was apparently calling "Chestnut Hill Farm" at that time), and both were immediately impressed. "The place is one of surpassing beauty," the writer gushed, "and a kinder host never extended a welcome to a friend, than the one who presides over the table of this elegant mansion." Welch's collection of trotters was heralded as being "among the finest in the country."[32] In the years to come, Flora Temple

FACING: Flora Temple's name was so ubiquitous by the 1850s that her image was routinely appropriated for advertising purposes. L. Prang and Company, tobacco labels, Boston, circa 1856. Courtesy of American Antiquarian Society.

Bull's Run, Bull's Run, Broadside, circa 1861. Courtesy of American Antiquarian Society.

would produce some of her own remarkable horses, including Lady Temple (1868) and Prince Imperial (1869).[33]

Over the preceding few years Welch also assembled an impressive collection of stallions, including Knight of Ashland, Brown Horse, and Jack Lane, and he was publicizing his faith "in the efficacy of a cross between the blooded horse and the trotter." So impressed were Welch's 1866 visitors that one writer concluded that "his stable is destined to become the most celebrated in the whole country."[34] By 1869, however, Welch was done with breeding trotters, turning his attention almost exclusively to thoroughbred racers. By 1872 John McCloskey was still the superintendent of the mares and stallions, but Welch had turned over responsibility for the thirty or so trotters, including old Flora Temple, to a Mr. Conover.[35] When Flora Temple died on December 21, 1877, she was so revered by Welch that he had her buried under a large granite slab within sight and just to the east of "Old" Erdenheim.[36]

John McCloskey's presence at Erdenheim may have been the most consistent element of the environment at Erdenheim between 1862 and the end of the century. His name appeared frequently alongside Welch's in the press, and it was clear that Welch often preferred to defer to McCloskey when answering questions about the horses. Born circa 1839 in Ireland, McCloskey married his wife, Anna, eight years his junior, sometime after his arrival at Erdenheim in either late 1862 or early 1863. They had at least four children by 1880. Correspondence from one of his sons, John F. McCloskey, indicates that McCloskey's brother or cousin, Edward McCloskey, also worked at Erdenheim as a gardener until 1880; Edward's daughter Henrietta and son Robert were named for members of the Welch family.[37]

Welch was not only a horseman, however. Fulfilling the expectations of a "gentleman farmer," he also invested heavily in Brahmin cattle, and his cross of his Brahmins with "ordinary cattle" was revered for producing "the most patient, hardy, and tractable of work cattle." Welch also earned kudos for his impressive flock of Damascus sheep.[38] Welch kept other animals at Erdenheim, too. John F. McCloskey, who was a young boy when Welch sold the farm in 1882, remembered that Welch at one point brought a herd of buffalo to Erdenheim and kept them with his West Indian cows on the lower field near Wissahickon Avenue.[39] Indeed Erdenheim under Welch appears to have been something of a menagerie. When a correspondent for the *Spirit of the Times* visited the stud farm of this "gentleman of rare enterprise and energy" in 1872, he found Welch's three young sons there to greet him. Their first priority was not to reveal Erdenheim's famous horses but instead to introduce the correspondent "to their pony from North Carolina, to the English mastiff bitch, the Cuban bloodhound, the great boar-hound from the banks of the Danube, the swan-like geese from the East Indies, and many other things of great interest, among which were the rows of nests in course of construction against the walls and under the eaves of the great stone barn, by those skilful architects and plasterers, the swallows."[40]

In addition to the stone stable that still stands on the east side of Stenton Avenue opposite "Old" Erdenheim, believed to have been built in 1870, Welch also made other improvements to the grounds at Erdenheim to facilitate his training of horses.[41] Map images from 1877 (reproduced earlier in this chapter) clearly show that Welch had a large racing oval on the premises, on the southeast corner of the intersection of West Mill Road and Township Line Road, covering the present location of the large wooden barn on the east side of Stenton Avenue, the house complex just north of that barn, and the large building to the north of that.[42] There is no evidence to suggest that Welch ever held races at Erdenheim,

FACING: 1870 Stable built by Aristides Welch. Photo by Eric Plaag, January 2012.

however, or that he encouraged the society set to descend en masse there, as later owners often did. This may indeed have been because an 1820 state law had proclaimed horse racing a "common nuisance" and the awarding of a purse in such a contest gambling (which was also illegal). An 1879 case before the Montgomery County court affirmed that horse racing was illegal in the county.[43] Among the other buildings on Welch's Erdenheim property was a house on the north corner of present-day Wissahickon Avenue and Stenton Avenue, previously occupied by Yeakle and Potts before Welch's purchase of the land there.[44] Six years later Welch apparently leased this property to Godfrey and Company.[45]

Leamington

Foaled in 1853, Leamington was perhaps the most famous stallion to ever stable at Erdenheim and was certainly the key to Welch's continuing success as a breeder. Leamington won eight of the twenty-four races he ran up to the age of six, often by massive margins. Purchased in 1866 and put out to stud by the Rawcliffe Stud Company, Leamington did not impress his British handlers. From there he went to America, purchased by R. W. Cameron of Staten Island, who in turn shipped him to General Abe Buford's Kentucky farm, where he was again a disappointment. A move to New York over the following three seasons didn't help much either, perhaps because he was paired with English mares quite similar to those he had failed with in England.[46]

Leamington's first crop of racers—Longfellow, Lyttleton, Enquirer, and Lynchburg—did catch the eye of Aristides Welch, though, and for this reason, Welch purchased Leamington in December 1869. There are some indications that the price of $11,000 was so outrageous and beyond Welch's means that he had to secure assistance from an unidentified wealthy Philadelphian to complete the purchase, but Welch was committed to making use of Leamington. After standing him elsewhere for three seasons, Welch brought Leamington to Erdenheim in 1871, where he quickly discovered that Lexington mares were the trick to making Leamington productive. What followed was a roster of champions that were the envy of every other American breeder: Hyder Ali, Parole (the leading purse winner in 1881, at $82,184), Sensation, Onondaga, and Aristides (first winner of the Kentucky Derby).[47] As he had done with Flora Temple, Welch again celebrated his noteworthy purchase by calling Edward Troye to Erdenheim to paint a commemorative image of Welch's new stud—an image dated October 21, 1871.[48]

Contrary to public perception, Leamington was not the monster that many in the press frequently made him out to be. Interviewed about the old horse in 1888, well after Leamington's passing, John McCloskey observed that "no horse could have been more kind." McCloskey then related that he often climbed into Leamington's stall, laid down in the hay, and allowed Leamington to "come up and rub his nose all over his (McCloskey's) face and lick it with his tongue. Like men of strong dislikes, he also had his likings, and to those he liked he was the most amiable and affectionate of horses."[49]

It is difficult to overstate the importance of Leamington to Welch's breeding operations or his iconic status in the horseracing world. In February 1873, when a correspondent from the *New York Herald* visited Welch's Erdenheim, the proprietor had taken to calling the place the "Leamington Stud Farm," after his prized stud, instead of "Chestnut Hill Stud Farm" (by which it was best known during the 1860s and 1870s) or "Erdenheim, or Happy Home" (the name Welch apparently used for the house itself). The correspondent suggested that his invitation

FACING: Leamington, from *Famous Horses of America,* 1877, Keeneland Library.

The Property of A. Welch, Esq., Chestnut Hill, Phila. LEAMINGTON.

to Welch's farm stemmed from a chance meeting with Welch on Chestnut Street in Philadelphia, but his arrival at the farm was accompanied by the news that Lyttleton, Welch's "hope and pride among his racing stallions, had been permanently ruined for racing purposes while exercising a few days previous." The correspondent then caught a glimpse of nineteen-year-old Leamington as the stable boy moved him into one of the stalls and waxed poetic about his astonishing appearance: "His color seems to be a rather rich purple as he stands in the stall; when brought into the light it is a dark brown. Sleek, long-bodied, clean, faultless limbs, unshod; head erect and eyes of fire, the smoke of his nostrils bore on its wings the breathings of his majestic nativity."[50]

Leamington's breeding successes continued when he was finally paired with Maggie B. B., a chestnut mare by Australian out of Madeline. James B. Clay Jr., a grandson of Henry Clay, named Maggie B. B. for his unrequited love, Miss Maggie B. Beck, the daughter of a U.S. congressman. Beck's father had disapproved of Clay's advances and arranged her marriage to another man. When she died two weeks after her marriage, Clay named the mare for her.[51] Tales of tortured romance aside, it was Maggie B. B.'s racing success against those early Leamington colts—Enquirer, Lynchburg, and Lyttleton—that attracted Welch's attention, and in 1871, while Leamington was still at New York, Welch bred her for the first time with his stallion. More champions followed, including Pera, Jaconet, Harold, Francesca, and Iroquois, the first American horse ever to win the Epsom Derby in England. When Leamington died from a heart attack in May 1878, his legacy was established; his ten years at stud had made him the leading sire in four of those years.[52] To quantify that success more specifically, from 1875 to 1881, Leamington's sons and daughters secured racing winnings totaling more than $437,000 for their owners.[53]

Welch enjoyed a fine reputation for the manner in which he raised and treated his horses. His 1890 obituary in the *Spirit of the Times* proclaimed him "the most successful breeder of his day," complimenting him and his methods. "He never nursed or coddled his yearlings or fattened them for appearance's sake," the blurb read, "but allowed them full range in a forty-acre paddock, which gave them fine legs and strong hearts."[54] He was also renowned for his innovations in horse feed. Writing more than a dozen years after Welch's death, one anonymous correspondent to the *Southern Planter* remembered that Welch had discovered that the use of corn stover mixed with hot water and other grains was excellent for promoting growth in colts: "This corn stover, meal, and bran feed was the one used by Mr. Aristides Welch at night. Every horse—Leamington, brood mares, and weanlings—were fed it every night. Oats and timothy morning and noon. . . . He saved every cornstalk and had it cut up by a machine and put in his barn."[55] Some speculated that the fine limestone rock undergirding Erdenheim was what made Welch's strong and nimble champions. Still others claimed that it was the manner in which he exercised his horses that brought him success. Visiting the 40-acre paddock (located adjacent to the stone barn on the east side of Stenton that still stands), in which Welch had exercised Leamington's offspring, a reporter in 1888 observed, "It was to this exercise obtained here . . . that many attributed the wonderful racing excellence of the Leamington youngsters, as they seldom had sore shins, and in the exercise developed the heart, lungs, and muscles, and in a measure prepared them for the trainers' hands."[56]

FACING: Maggie B. B., from *Spirit of the Times,* March 1, 1884, Keeneland Library.

MAGGIE B. B.,
QUEEN OF THE ERDENHEIM STUD.

7
1

Welch's Other Champions

When news broke that Pierre Lorillard's colt Iroquois, bred of Leamington and Maggie B. B., won the English Derby, the consequences may have been better for Welch than for Lorillard. Already renowned for his breeding innovations, Welch's enormous popularity soared higher, and the press sought to interview him at every turn. Welch had sold Iroquois to the Lorillard brothers two years earlier, in 1879, for "about $1,000," as part of his annual sale of yearlings (which also included about a dozen other horses, including Spinaway, Saunterer, and Blazes).[57] Indeed so heartbroken was Welch following the death of Leamington that he thereafter had the auctioneer stand on Leamington's grave in the yard near his home, which certainly must have had a dramatic effect and served as a reminder of how limited the remaining crop of Leamington's immediate descendants was.[58] Iroquois was one of the last of Leamington's offspring, and as Welch once told the *Spirit of the Times,* he was the best colt Welch ever bred. Iroquois became so popular in the weeks following his victory that advertisements offering a lithograph appeared for months afterward in the sporting papers.[59]

Contrary to public sentiment, however, Iroquois was not believed by either Welch or McCloskey to be the most promising yearling among the crop in 1879. Interviewed in 1888, McCloskey called Iroquois "the most arrogant, domineering youngster that he ever had at Chestnut Hill," worse even than Harold, the kind of horse that, out of "pure cussedness," would go among the other yearlings with a cornstalk in his mouth, daring the others to challenge him for it. Sensation, McCloskey insisted, not Iroquois, was "by long odds the grandest mover that ever galloped over the Erdenheim paddocks."[60]

FACING: Iroquois at the Epsom Derby, from *Spirit of the Times,* July 2, 1881, Keeneland Library.

Ultimately the record seems to suggest that Iroquois was not the same after his accomplishments in England. This may, in part, have been a result of a string of broken blood vessel injuries he suffered in 1882, when he could not race, and several unimpressive finishes thereafter in 1883. Once he was back in America, nothing seemed to go well, the blood vessel injuries continued, and he was retired to stud. As Fred Archer, the English jockey who led him to victories at Epsom and St. Leger, said at the time, "They shouldn't have raced him here. He was in no condition when he left England to race over a mile. He was the most resolute sort of a colt, and ran as though he liked racing."[61]

In addition to Iroquois, who may have been the most famous horse bred by Welch, Erdenheim produced other champions. Rysdyk (foaled 1865) and Strathmore (foaled 1866) were two trotters who in turn produced their own champions, most notably Strathmore, whose descendants produced between 1874 and 1950 included numerous record-breaking horses, including a total of thirty-five Standard pacers, believed to be the highest number produced by any son of the famed Hambletonian. Nevertheless Welch was at first infuriated that his trotting sire and dam had produced a pacer rather than a trotter, and when both breeding and trotting with Strathmore (then known as Goodwin Watson) proved impossible, Welch used him as a stable pony. By 1873 Welch felt fortunate to pass him off for a small price to R. G. Stoner, who promptly turned Strathmore into a low-end sire and realized Strathmore's capabilities only after 1879. While one of Strathmore's foals at Erdenheim, Chestnut Hill, eventually went on to be a champion and record-breaker, Welch never knew until it was too late what he had let get away.[62]

In addition to the most famous horses of Erdenheim, Welch bred countless others that went on to fame under the ownership of other men. Among them were Harold, winner of the 1879 Preakness for George Lorillard; the aforementioned Saunterer, who won at the Preakness and Belmont just two years later; and Panique, whose 1884 victory at Belmont came after Welch had moved on from Erdenheim.[63] Nevertheless the success of these champions created a high demand for Welch's ponies, both at the annual yearling sales and for individual horses that had already proven themselves. At one point the desire for Erdenheim's champions was so great that Welch refused an offer of $15,000 for Harold, the "broken down" brother of Iroquois—a staggering sum for the day.[64] On another occasion, in 1870, he took $31,000 for Lady Thorn—at the time a record price for a horse.[65]

By 1872 Welch's reputation as a breeder was sufficiently established that he more often served as a supplier of champions to other owners than as a backer of racers himself. One of Welch's frequent buyers was Pierre Lorillard of New York, perhaps the most dominant owner of racers during this period, who on occasion would descend on Erdenheim for the annual sale of Erdenheim yearlings and buy nearly the entire crop. Among Lorillard's purchases was Parole, a brown gelding by Leamington and Maiden, who was foaled by Welch at Erdenheim in 1873 and promptly sold to Lorillard. Parole won his first race—the July Stakes at Long Branch in 1875—and his second race, at the Augusta Stakes, ensuring his owner that the purchase had been a wise one. Indeed, out of his first six races, Parole won four of them.[66] For his career, he won fifty-nine of the 137 races he ran, winning total purses of $82,909.25. He was renowned for "lying away and coming with a burst of speed at the finish," and it was said that no other horse of the period was better down the stretch.[67]

Welch's fame culminated in a charming gesture by owner H. P. McGrath after he purchased one of Welch's yearlings foaled of Leamington and Sarony in 1872. As a measure of his gratitude to Welch, McGrath named the horse he had purchased "Aristides." His first full season, 1874, was not entirely successful, although he finished strong, and 1875 didn't hold out much promise either, until he won the inaugural Kentucky Derby at Louisville—a mile and a half race—in just over 2:37, beating out his greatest rival, Ten Broeck. A month later he won the Withers Stakes at Jerome Park, then the Jerome Stakes in early October and the Breckenridge Stakes at Baltimore in late October. The next year he set a record time at Lexington in the four-year-old Sweepstakes, solidifying his reputation as an especially accomplished long-distance racer.[68]

The Sale of Erdenheim

By December 1881 Welch's passion for the industry seemed to be waning, even if he was at the pinnacle of success and still regarded as resembling "a hale, bluff man of 50" in spite of being seventy years old.[69] When news broke on January 8, 1882, that Welch had decided to sell Erdenheim, though, much of the racing world was stunned, primarily by his decision to sell it for about two-thirds of its estimated $150,000 value to Commodore Norman W. Kittson of Minnesota. While it is true that the farm had been on the market for nearly a year, many observers interpreted the move as a desperate effort to raise cash because of some sort of embarrassing financial difficulty. In addition to the 200 acres at Erdenheim, the sale was to include "all its stock, comprising over 80 horses, mares, and colts,

FACING: Parole, bred by Aristides Welch for Pierre Lorillard. From *Famous Horses of America,* Keeneland Library.

The Property of Pierre Lorillard, Esq., New York. PAROLE.

ARISTIDES.

THE PROPERTY OF H. P. M'GRATH, ESQ.

among which are the well-known stallions Alarm and Reform, and Maggie B. B., the dam of Iroquois." To many, such a bargain seemed preposterous.[70]

Some attributed Welch's decision to sell to his heartbreak over the loss of Leamington, "whose victories he never tired of describing to a circle of friends, among whom were . . . Col. John W. Forney, Sylvester J. Megargee, [and] August Belmont."[71] Others, however, knew better. Colonel S. D. Bruce, editor of *Turf, Field, and Farm,* offered a more plausible explanation: "It is true that Mr. Welsh [*sic*] has been anxious to dispose of the Erdenheim stud. During last Summer he offered it to me for $125,000. But his reasons for selling are not by any means the result of financial entanglements. They are of a domestic character purely. It is stated that he proposes to go to Europe and take his sons with him. Mr. Welsh has made money at Erdenheim. I have reason to believe that he has cleared at least 30 per cent on his investment."[72]

Welch's desire to flee the country may indeed have had a great deal to do with the exploits of his young sons. In August 1880, for example, his son William A. Welch was arrested and held on $600 bail after he pelted Georgie Parker, an actress who had jilted him, with rotten eggs during a performance at the Walnut Street Theatre. Welch subsequently failed to appear for his arraignment, although Parker seemed more interested in pressing charges against Rose Bunger, the matron of a boarding house where Parker had once lived, than Welch, whom Parker claimed to barely know.[73] As if that weren't embarrassing enough, Welch's older son, Robert A. Welch, a "scapegrace" figure whose exploits in the early 1880s were a perennial embarrassment to his father, soon found trouble of his own. In November 1881 son Robert forged an $800 check, cashed it, and asked a friend named "Petrie" to hold it for him. When the two traveled to New York and Robert demanded his funds from Petrie, proprietor of the Gotham Saloon at 26 Carmine Street, the latter refused, and Welch shot him in the head and chest, wounding Petrie severely. Petrie survived, but paying for his son's defense cost Aristides Welch more than $6,000. In late 1882 Robert Welch again went on a forgery spree, leading to his arrest in January 1883. In that case Robert defrauded his father's close friend and fellow breeder, Adam R. Magraw, by encouraging the latter to endorse a check on Welch's behalf in the amount of $675. With money in hand, Welch fled Magraw's home in Cecil County, Maryland (where he was to have been hunting with Magraw), and went on a duck-hunting spree in North Carolina. He was arrested when he returned to a boarding house at 248 Wyoming Street in West Philadelphia.[74]

FACING: Aristides, bred by Aristides Welch and named for his breeder, became the first winner of the Kentucky Derby. From *Famous Horses of America,* Keeneland Library.

Whatever Welch's reasons for selling Erdenheim, prospective buyers initially found Welch's various asking prices of $130,000 and $125,000 to be especially steep, considering the cheapness of land in Kentucky. "I could see no money in it," F. T. Walton offered. "I found it too valuable. . . . I discovered it was pretty high-priced land."[75] Such pressures, perhaps coupled with the embarrassment over his son's shenanigans, forced Welch to offer Kittson a bargain, which Kittson gladly accepted. Welch apparently thought better of his sale, though, after his children persuaded him that "he would be discontented away from his old home." By then son James had either left home or was away at school, but Robert, a medical student, and William, a law student, still lived close at hand. Erdenheim also supported a staff of fourteen employees who lived on site, nearly all of them from Ireland, including several domestic workers and a stableman, a coachman, a watchman, and a "cow-man." John McCloskey, Welch's faithful stableman, by then had his own residence near Erdenheim, where he lived with his wife and four children.[76] Welch pleaded with Kittson several days later to reverse

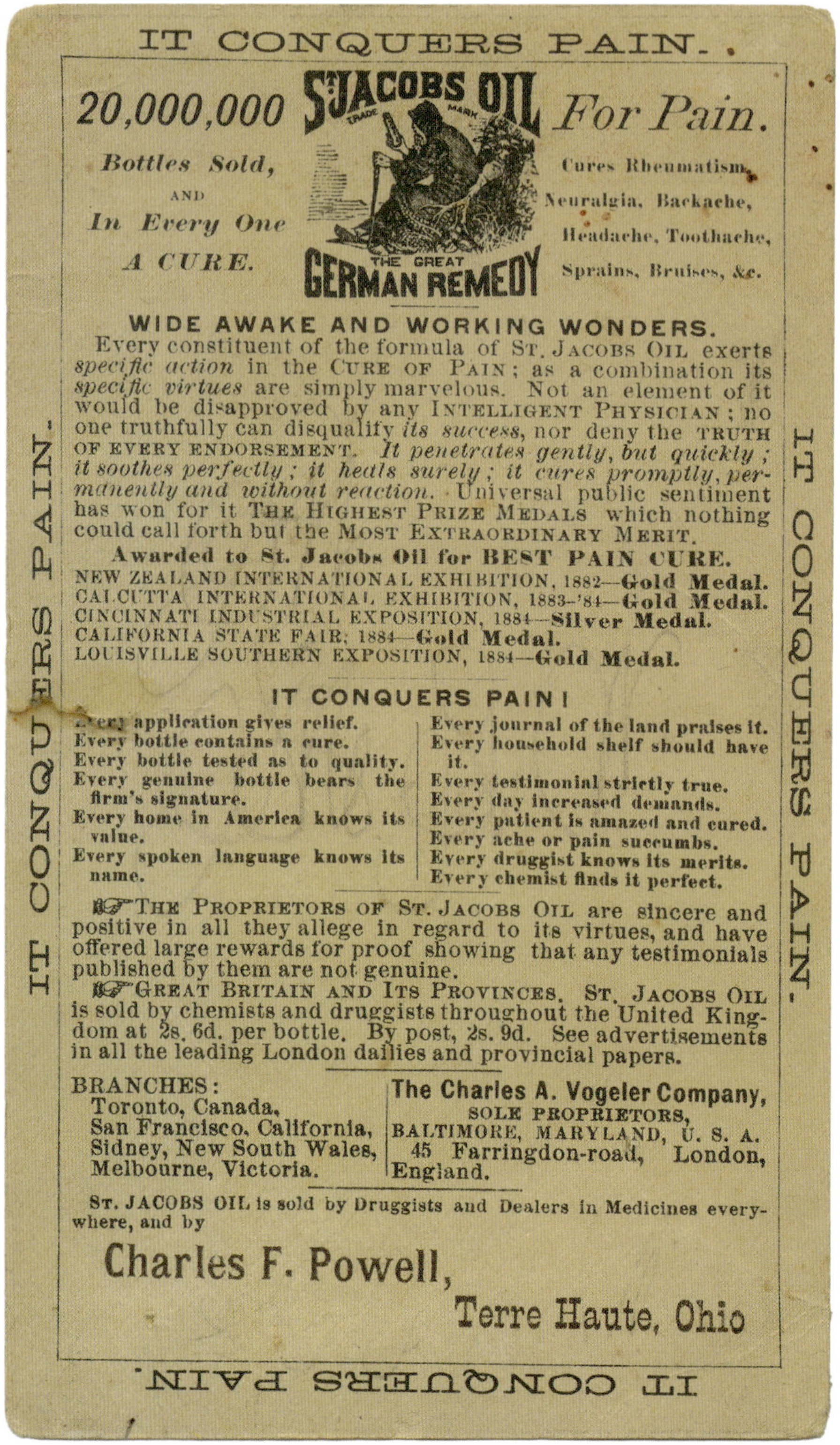

IT CONQUERS PAIN.

20,000,000 *Bottles Sold,* AND *In Every One A CURE.*

ST. JACOBS OIL TRADE MARK THE GREAT GERMAN REMEDY

For Pain. Cures Rheumatism, Neuralgia, Backache, Headache, Toothache, Sprains, Bruises, &c.

WIDE AWAKE AND WORKING WONDERS.

Every constituent of the formula of ST. JACOBS OIL exerts *specific action* in the CURE OF PAIN; as a combination its *specific virtues* are simply marvelous. Not an element of it would be disapproved by any INTELLIGENT PHYSICIAN; no one truthfully can disqualify *its success*, nor deny the TRUTH OF EVERY ENDORSEMENT. *It penetrates gently, but quickly; it soothes perfectly; it heals surely; it cures promptly, permanently and without reaction.* Universal public sentiment has won for it THE HIGHEST PRIZE MEDALS which nothing could call forth but the MOST EXTRAORDINARY MERIT.

Awarded to St. Jacobs Oil for BEST PAIN CURE.

NEW ZEALAND INTERNATIONAL EXHIBITION, 1882—**Gold Medal.**
CALCUTTA INTERNATIONAL EXHIBITION, 1883-'84—**Gold Medal.**
CINCINNATI INDUSTRIAL EXPOSITION, 1884—**Silver Medal.**
CALIFORNIA STATE FAIR; 1884—**Gold Medal.**
LOUISVILLE SOUTHERN EXPOSITION, 1884—**Gold Medal.**

IT CONQUERS PAIN!

Every application gives relief.
Every bottle contains a cure.
Every bottle tested as to quality.
Every genuine bottle bears the firm's signature.
Every home in America knows its value.
Every spoken language knows its name.
Every journal of the land praises it.
Every household shelf should have it.
Every testimonial strictly true.
Every day increased demands.
Every patient is amazed and cured.
Every ache or pain succumbs.
Every druggist knows its merits.
Every chemist finds it perfect.

☞THE PROPRIETORS OF ST. JACOBS OIL are sincere and positive in all they allege in regard to its virtues, and have offered large rewards for proof showing that any testimonials published by them are not genuine.

☞GREAT BRITAIN AND ITS PROVINCES. ST. JACOBS OIL is sold by chemists and druggists throughout the United Kingdom at 2s. 6d. per bottle. By post, 2s. 9d. See advertisements in all the leading London dailies and provincial papers.

BRANCHES:
Toronto, Canada,
San Francisco, California,
Sidney, New South Wales,
Melbourne, Victoria.

The Charles A. Vogeler Company,
SOLE PROPRIETORS,
BALTIMORE, MARYLAND, U. S. A.
45 Farringdon-road, London, England.

ST. JACOBS OIL is sold by Druggists and Dealers in Medicines everywhere, and by

Charles F. Powell,
Terre Haute, Ohio

IT CONQUERS PAIN.

IT CONQUERS PAIN.

IT CONQUERS PAIN.

the sale for the forfeiture of $2,500 and two brood mares. Kittson agreed, but by May 1882, the two men reopened negotiations, agreeing on a price of $125,000, to account for the presence of approximately twenty-five more horses on the farm.[77]

Welch's retirement after the sale of Erdenheim resulted in a life of isolation, no doubt contributed to by the disastrous events involving his sons. When he died in 1890 at the age of seventy-nine, his obituary in the *Philadelphia Inquirer* suggested a bit about who the man was: "Beyond his fame as a raiser of many of the most noted celebrities of the turf, Mr. Welch lived a secluded life, and since his retirement . . . he had been seen but little in the active world."[78] Other publications were not so kind. Announcing his death at the Bingham House in Philadelphia on April 9, 1890, *Turf, Field, and Farm* observed that Welch possessed "a bluntness of speech which told against his popularity," a trait the paper claimed he had picked up while in the navy.[79] That bluntness, or at least a latent bigotry, had certainly surfaced from time to time in interviews he granted, as when he told one reporter in 1881, "Colored trainers are calculated to wreck horses. They have neither the judgment nor the sense for the responsibilities of such work."[80]

But there was also a darkness that seemed to haunt Welch, especially in his later years. "Shadows fell across the path of Mr. Welch in his closing years," the *Turf, Field, and Farm* death notice continued, "and he was not seen much in the places that once knew him. He and George Wilkes [the former editor and owner of the *Spirit of the Times*] were very friendly at one time, but a gulf separated them before Mr. Wilkes died."[81] Regardless of this characterization of their relationship, Welch must not have been too miffed at Wilkes. For years Welch had been investing in both the *Spirit of the Times* and the *Sportsman,* and by the time of Wilkes's death in 1885, Welch owned all of Wilkes's former shares in the paper.

FACING: St. Jacobs Oil trade card, circa 1885, author's private collection.

As an indication of his fondness for Wilkes, Welch gave the entirety of his interest in the paper to Wilkes's children following Wilkes's death.[82]

Welch's irritation with the media of his day, and any perceived rift with Wilkes and the *Spirit of the Times* in particular, might be understandable, though, considering the ways in which the newspapers frequently appropriated Welch's name for increased advertising revenue. One of the most frequently reprinted articles about him shortly before his retirement focused on a line in an interview he gave at Erdenheim to the Philadelphia correspondent from the *New York Sportsman,* in which Welch allegedly endorsed the snake oil potion widely marketed as St. Jacobs Oil, calling it "an excellent thing for horses." The British makers of St. Jacobs Oil claimed that it was "the great German remedy" for a host of complaints, including rheumatism, "quinzy" (tonsillitis), soreness of the chest, frosted feet and ears, lumbago, and gout. Given that it apparently contained 83 percent turpentine with traces of camphor, 10 percent ether, 5 percent alcohol, and 2 percent carbolic acid (with only small amounts of aconite and capsicum), it was likely to cause renal failure and a host of other problems if taken internally or applied topically. In light of Welch's reputation for careful, studied treatment of his horses and consideration of the circumstances of their breeding and care, it seems unlikely that a former seamen like Welch, who would have known of the properties and variant medicinal uses—and consequences—of turpentine, would have actually endorsed such a product. At a cost to its makers of only two pence for a bottle that sold in Great Britain for two shillings, six pence, though, his endorsement, whether genuine or not, certainly must have been a boon for sales and profits.[83]

A careful examination of the origins of his endorsement reveals a possible source of Welch's irritation with Wilkes. In the original interview, Welch hesitated when pressed about the use of St. Jacobs Oil, deferring instead to his head groom: “Well, John [McCloskey] here knows more about the use of liniments than I do, as he takes direct charge of my stock, and is an experienced man, but I have heard a good deal about St. Jacobs Oil, and will give it a good trial. I believe it to be an excellent thing, and John will use some of it the first opportunity, but we have very little sickness of any kind among my stock just now.” Welch then deferred further, citing the recommendations of other breeders who had used the product and encouraging the reporter to trust their word rather than his. When the reporter returned a number of days later and confronted McCloskey and Welch about St. Jacobs Oil again, Welch deferred once more to McCloskey, who replied, “I have had a chance to try it. Come along with me, and I will show you a colt I rubbed with it.” Pointing out the colt, McCloskey added, “When I first used St. Jacobs Oil on that colt he jumped about in the most violent manner, and was for a time almost unmanageable. I did not think that was a bad sign, for it only proved to me that there was strength in the liniment.” After further consideration, and an explanation that he had applied St. Jacobs again to the colt with favorable results, McCloskey added, “My use of the Oil on that colt has proved to me that it is an admirable preparation for horses, and I shall continue to use it.” When asked whether he had used it on any other horses, McCloskey again explained that their horses were rarely injured, and that St. Jacobs Oil might be better suited to work horses, such as those driving trolley cars. Unsatisfied, the reporter pressed Welch again, to which Welch said, “I want John to keep on using that oil, for I think it is a very good thing. . . . By the next time you come out to see me I think I will have something more to say about St. Jacobs Oil, if you feel any interest in hearing anything about it.”[84]

Over the ensuing months, however, this lengthy interview with McCloskey and Welch, which had been almost entirely focused on St. Jacobs Oil, began to appear in other newspapers in a greatly condensed form. None of McCloskey’s qualifications or Welch’s deferrals appeared in the brief blurbs that followed in other newspapers, which instead touted Welch’s endorsement. These condensed versions referenced Welch’s unfortunate and heavily qualified observation that the liniment was “an excellent thing” completely out of context. Perhaps the most distressing version, however, was a longer excerpt that appeared in *Spirit of the Times* the following spring. Here most of Welch’s deferrals and hedges were excised, but his commentary about the favorable things he had heard and his desire to see the product tested was retained. The net result was a rather lengthy article that seemed to be an all-out endorsement of St. Jacobs Oil by Welch and offered little in the way of actual news about Welch or Erdenheim.[85]

By the time these articles first began appearing in the press in late 1881, it was well known that Aristides Welch was already irritated with the papers and their distortions. Following the news of Iroquois’s victory at Epsom, for example, reporters struggled to pin down either Pierre Lorillard or Aristides Welch, the latter of whom, the *Spirit of the Times* reported, “denounces the interviewers, and declares that they fabricated interviews with him which never occurred. We had our attention called to one of these which appeared in the columns of a Philadelphia contemporary, and while we cannot for a moment allow ourselves to doubt Mr. Welch’s word, we must give the writer of this the credit of ‘knowing this man,’ for he dotted every i and crossed every t to harmonize the article with the hearty manner and impromptu style of the hospitable proprietor of Erdenheim.”[86] There seems little doubt that the “Philadelphia contemporary” was the

FACING: The site of “Old” Erdenheim, behind the lamppost and fence, with granite slab graves of Leamington and Flora Temple in the foreground. Photo by Eric Plaag, May 2012.

same Philadelphia correspondent from the *New York Sportsman,* but whether or not the original interview appeared before June 1881 is not clear.

As for the physical signs of Welch's legacy at Erdenheim, there may not be a better emblem than the series of granite slabs lying just to the east of "Old" Erdenheim's location. One *Spirit of the Times* writer described the scene perfectly in 1888, immediately following the death of Erdenheim's next owner, Commodore Norman W. Kittson:

> Two great granite slabs on the lawn under the magnolia trees to the east of the mansion, simply inscribed, one "FLORA TEMPLE," the other "LEAMINGTON," serve to remind us of the greatest American trotting mare of her day and of the most successful stallion that ever crossed the seas, who lies beneath the stone close to the scene where once he held court and gave the world an Iroquois, a Parole, a Sensation, a Spinaway, a Harold, and an Onondaga. The tomb of Leamington might serve as a shrine for the racing devotee to worship. It was covered with fallen leaves, and aroused a flood of reflection.[87]

Those stone slabs, now joined by a third one for Maggie B. B., still grace the lawn of "Old" Erdenheim, even though no signs remain of that grand residence in which Welch, Blight, Hocker, and so many others lived for more than 150 years, just a short walk from that fabled resting place. Indeed those old tombstones and the stone walls of the old stable on the east side of Stenton may be the only surviving reminders of the breeding operation and the fame that Welch brought to Erdenheim.

Notes

1. This date is from Welch's grave marker, located at Oak Hill Cemetery in Washington, D.C. See also *Genealogies of Pennsylvania Families from the Pennsylvania Genealogical Magazine,* volume 1: *Arnold-Hertzel,* 784, Ancestry.com. This latter source incorrectly reports Welch's year of birth as 1817 but correctly indicates his place of birth as Shippensburg. In two separate passport applications, Welch's son James A. Welch listed his father's place of birth as St. Louis, Missouri, for reasons that are not clear. See James A. Welch passport application, June 21, 1919, *Passport Applications, January 2, 1906–March 31, 1925,* NARA Series M1490, roll 807, National Archives and Records Administration, and James A. Welch passport application, November 22, 1920, *Passport Applications, January 2, 1906–March 31, 1925,* NARA Series M1490, roll 1418, National Archives and Records Administration, Ancestry.com.
2. *Washington, DC, Marriages, 1826–50,* Washington, D.C., District Court, Ancestry.com. Welch's name is listed as "Welsh" in this document. Welch spent a significant amount of time in California during the early 1850s, prior to California's statehood and during his service in the U.S. Navy, and he may have had a home there during this time. Neither Welch's nor Sidney Francis's name appears in the 1850 U.S. Federal Census.
3. John Henry Sherburne, *The Life and Character of John Paul Jones, a Captain in the United States Navy* (New York: Adriance and Sherman, 1851), 393, Making of America Collection; General Cadmus M. Wilcox, *History of the Mexican War,* ed. by his niece Mary Rachel Wilcox (Washington, D.C.: Church News Publishing, 1892), 701, Making of America Collection; *Senate Executive Journal,* March 11, 1856, 62, American Memory Collection. Welch was still listed among the crew of the *Merrimac* when it sailed in August 1856, but it is almost certain that he resigned in February 1856, per copies of original documents in possession of Mary Jane Echenhofer and reviewed by the author. See "Naval Intelligence," *New York Herald,* August 16, 1856. The infamous ironclad *CSS Virginia* was later built on her hull. The *Princeton,* meanwhile, was infamous for an incident that occurred in 1844, before Welch's service aboard her. On February 28, on the Potomac River, a massive explosion of the "Peacemaker," one of the steamer's twelve-inch muzzle-loaded guns, killed several of the four hundred dignitaries on hand for a demonstration of the ship's capabilities. Among the dead were Secretary of State Abel P. Upshur, Secretary of the Navy Thomas Gilmer, and six others, including President John Tyler's valet. The president and the rest of his cabinet, all of whom were on board, narrowly escaped death. The ship was broken up in 1849, when her timbers were found to be rotten. For more on the *Princeton,* see Robert W. Merry, *A Country of Vast Designs: James K. Polk, the Mexican War, and the Conquest of the American Continent* (New York: Simon and Schuster, 2009), 65–66, and U.S. Naval Historical Command, "Princeton," *Dictionary of American Naval Fighting Ships,* available at http://www.history.navy.mil/danfs/p12/princeton-i.htm.
4. "Marriage," *Daily National Intelligencer,* March 6, 1856.
5. "Our Washington Correspondence," *Daily Globe* (San Francisco), November 6, 1857; "Land Sales in Albemarle," *Alexandria Gazette,* September 28, 1857. Blenheim was listed in the National Register of Historic Places in 1976. The official nomination document does not mention the Welches or any other owners after Stevenson.
6. Mary Jane Echenhofer to Eric Plaag, August 20, 2012, private email communication.

7. *Dewey v. Lambier* (1857), 7 Cal. 347.
8. "Letter from Washington," *San Francisco Bulletin,* May 17, 1861. On the 1860 Census, see "Lancaster, North East Ward, Lancaster, Pennsylvania," *1860 United States Federal Census,* roll M653-1126, National Archives and Records Administration, Ancestry.com. Welch is inexplicably listed as "Adam Welch" and as being forty-five years old (he was forty-nine), while Henrietta is listed as "Henrietta K. J. Welch," born in Tennessee, and thirty-three years old. The other data seems to make it certain that this was the Welch family, however, and neither Aristides nor Henrietta appears in the 1860 Census for any other state.
9. Atherton Blight to Aristides Welch, October 22, 1862, Deed Book 128, page 149, Historical Society of Montgomery County.
10. *1860 United States Federal Census, Population Schedule,* NARA microfilm publication M653, roll 1159, page 25, National Archives and Records Administration, available at ancestry.com.
11. On the first transaction, which was located just northeast of present day Stenton Avenue in Springfield Township, see Caleb Heydrick [Heirs of] to Aristides Welch, April 1, 1867, Deed Book 149, page 289, and Aristides Welch to Richard C. McMourtrie, November 21, 1872, Deed Book 205, page 262, Historical Society of Montgomery County. On the latter transaction, see Henry Harner to Aristides Welch, April 6, 1872, Deed Book 197, page 248, Historical Society of Montgomery County.
12. See Hiram Potts, et al., to Aristides Welch, January 19, 1863, Deed Book 28, page 370, and Sheriff's Sale, Henry A. Piper to Aristides Welch, May 28, 1870, Deed Book 183, page 3, Historical Society of Montgomery County.
13. *1870 United States Federal Census,* NARA microfilm publication M593, roll 1379, page 444B, National Archives and Records Administration, Ancestry.com, and James A. Welch Passport applications, June 21, 1919, and November 22, 1920. McCloskey was listed as twenty-five in the 1870 Census, and his name was spelled "McLusky." "Estates of the Dead," *Philadelphia Inquirer,* April 16, 1890, confirms the middle initials of each of Welch's sons.
14. "Mrs. Aristides Welch, Burial," Presbyterian Church of Chestnut Hill, *Historic Pennsylvania Church and Town Records,* reel 633, Historical Society of Pennsylvania, available at Ancestry.com.
15. "Chestnut Hill," *Turf, Field, and Farm,* January 6, 1866, 8.
16. J. Bennett Nolan, ed., *Southeastern Pennsylvania,* volume 2 (Philadelphia: Lewis Historical Publishing, 1943), 754.
17. John Wien Forney, *Anecdotes of Public Men* (New York: Harper Brothers, 1873), 419–21. The building housing Harry Connelly's was demolished in 1859.
18. Nolan, *Southeastern Pennsylvania,* 754. In a curious coincidence, Dr. James A. McCrea, the prior owner of Erdenheim during the early 1850s, invited President Buchanan to be his guest at a meeting of the Saturday Evening Club at his home at 2004 West Delancey Place in Philadelphia in January 1865. From their correspondence, however, there is little reason to believe that Buchanan visited Erdenheim during McCrea's ownership. See Dr. James A. McCrea to James Buchanan, January 9, 1865, James Buchanan Papers, section 91, box 43, folder 8, Historical Society of Pennsylvania.
19. Edward W. Hocker, "Montgomery County History," *Bulletin of the Historical Society of Montgomery County, Pennsylvania,* volume 12, Fall 1959-Spring 1961, 220. One later source makes the absurd claim that a total of eight Presidents stayed at Erdenheim before 1907. See "Where Robert N. Carson's Millions Will Rear a College for Orphan Girls," *Public Ledger,* October 24, 1907.
20. Much of the overview here on Davis's imprisonment and the efforts to release him is based on William J. Cooper Jr., *Jefferson Davis, American* (New York: Alfred A. Knopf, 2000), 560–68. See also, though, Lynda Lasswell Crist, ed., *The Papers of Jefferson Davis,* volume 12 (Baton Rouge: Louisiana State University Press, 2008), 198.
21. Cooper, *Jefferson Davis,* 564–68, 582.
22. Crist, ed., *Papers,* 117–18, 157, 299.
23. Samuel Yeakle, "'Erdenheim' and John George Hocker," 113. Yeakle attributes the story to Louis Kittson, the son of Erdenheim's owner during the 1880s. Edward Hocker, writing as "Norris" in the *Norristown Times Herald* at an unknown date, suggested that Yeakle had actually spoken with two men who personally saw Grant at Erdenheim. See Norris [Edward Hocker], "A Lafayette Tradition about Erdenheim," *Norristown Times Herald,* undated clipping in the Edward Hocker Scrapbooks, Historical Society of Montgomery County.
24. Nolan, *Southeastern Pennsylvania,* 754.
25. Terry Conway, "Leamington: 19th-Century Ruler," *Blood-Horse,* June 19, 2010, 1859.
26. "The Man on the Corner," clipping from unknown newspaper, September 23, 1909, Springfield Township Historical Society, 2008.96.2.
27. "Erdenheim Stock Farm," *Philadelphia Record,* unknown date (c. 1903), clipping in the personal archives of Mary Jane Echenhofer.
28. Henry T. Coates, "Short History of the American Trotting Horse," *Every Horse Owner's Cyclopedia,* edited by L. C. Bruce (Philadelphia: Porter and Coates, 1880), 523. For more on Flora Temple, see John Hervey, *The American Trotter* (New York: Coward-McCann, 1974), 455–57.
29. Coates, "Short History," 520–23; "Erdenheim," *St. Paul Globe,* August 24, 1896.
30. Coates, "Short History," 520–23; "Erdenheim," *St. Paul Globe,* August 24, 1896; Alexander Mackay-Smith, *The Race Horses of America: 1832–1872* (Saratoga Springs, N.Y.: National Museum of Racing, 1981), 305.
31. Mackay-Smith, *Race Horses,* 307.
32. "Chestnut Hill," *Turf, Field and Farm,* January 6, 1866, 8.
33. "Mr. Welch's Breeding Farm," *The Spirit of the Times,* November 13, 1869, 202. This article provides a lengthy list and description of many of the horses in Welch's stable at the time.
34. "Chestnut Hill," *Turf, Field and Farm,* January 6, 1866, 8.
35. "Mr. Welch's Stud Farm," *Spirit of the Times,* June 1, 1872; "Erdenheim Stock Farm," *Philadelphia Record,* unknown date (c. 1903), clipping in the private archival collection of Mary Jane Echenhofer.
36. "Erdenheim," *St. Paul Globe,* August 24, 1896.
37. Dr. John F. McCloskey to Mrs. Frank W. Echenhofer, April 20, 1948, Mary Jane Echenhofer

private archives; Whitemarsh, Montgomery County, Pennsylvania, *1880 United States Federal Census,* roll 1159, 240C, National Archives and Records Administration, Ancestry.com.

38. "Chestnut Hill," *Turf, Field, and Farm,* January 6, 1866, 8.
39. Dr. John F. McCloskey to Mrs. Frank W. Echenhofer, April 20, 1948, Mary Jane Echenhofer private archives.
40. "Mr. Welch's Stud Farm," *Spirit of the Times,* June 1, 1872.
41. Conway, "Leamington," 1859.
42. See G. M. Hopkins, *Atlas of Philadelphia and Environs,* 1877, plate 67.
43. Edward W. Hocker, "Montgomery County History," 220.
44. This house is identified as Yeakle and Potts's on D. J. Lake and S. N. Beers, *Map of the Vicinity of Philadelphia* (Philadelphia: Stone and Pomeroy, 1860) and Welch's on *Atlas of the County of Montgomery and the State of Pennsylvania* (Philadelphia: G. M. Hopkins, 1871).
45. *Combination Atlas Map of Montgomery County,* Pennsylvania (Philadelphia: J. D. Scott, 1877), plate 90.
46. Susan Rhodemyre, "Aristides of Pennsylvania," *Thoroughbred Record,* January 23, 1980, 286–87.
47. Rhodemyre, "Aristides," 286–87, and "Erdenheim," *St. Paul Globe,* August 24, 1896.
48. Mackay-Smith, *Race Horses,* 335. There is some evidence to suggest that Troye was at Erdenheim in July 1869 to paint a young Prince Imperial with his older sister Kitty Temple, but this painting does not appear to survive. See Mackay-Smith, *Race Horses,* 326.
49. "Post and Paddock," *Spirit of the Times,* November 10, 1888, 589.
50. "Chestnut Hill Stud Farm," *New York Herald,* February 22, 1873, 5.
51. "Chestnut Hill Stud Farm," *New York Herald,* February 22, 1873, 5; "The English Derby," *St. Paul Daily Globe,* June 6, 1881; and Rhodemyre, "Aristides," 287. On Maggie B. B.'s lineage, see "Mr. Welch's Stud Farm," *Spirit of the Times,* June 1, 1872.
52. Rhodemyre, "Aristides," 287; "Reminiscences of the Turf: Imp. Leamington," *Thoroughbred Record,* November 5, 1904, 295.
53. "Com. Kittson's Latest," *St. Paul Daily Globe,* January 13, 1882, 2.
54. Untitled obituary notice, *Spirit of the Times,* April 12, 1890, 517.
55. V. of Tennessee, "The Horse: Feeding Thoroughbred Yearlings," *Southern Planter,* August 1900, 456.
56. Untitled column, *Spirit of the Times,* November 3, 1888, 560.
57. "The English Derby," *St. Paul Daily Globe* (Minnesota), June 6, 1881, and "History of Iroquois: A Good Purchase," *Daily Inter Ocean,* June 2, 1881, 1.
58. "Horse Dots," *Louisiana Democrat,* August 7, 1878.
59. Untitled column, *Spirit of the Times,* June 11, 1881, 500.
60. "Post and Paddock," *Spirit of the Times,* November 10, 1888, 589.
61. W. S. Vosburgh, *Racing in America, 1866–1921* (New York: Scribner's Press, privately printed, 1922), 122–23.
62. Hervey, *American Trotter,* 151–58.
63. Susan Rhodemyre, "Aristides of Pennsylvania," *Thoroughbred Record,* January 23, 1980, 286.
64. "Death of Aristides Welch," *Philadelphia Inquirer,* April 10, 1890.
65. Hocker, "Montgomery County History," 220.
66. *Famous Horses of America* (Philadelphia: Porter and Coates, 1877), 61.
67. Vosburgh, *Racing in America, 1866–1921,* 102–3.
68. *Famous Horses of America,* 45.
69. "Aristides Welch: The Breeder of Iroquois and Parole," *Sacramento Daily Record Union,* December 13, 1881.
70. "The Sale of Erdenheim: Aristides Welch's Stock Farm Sold at a Sacrifice," *New York Times,* January 9, 1882.
71. "Flames at Erdenheim," *New York Times,* October 1, 1884.
72. "The Sale of Erdenheim."
73. "An Act Not in the Bills," *Philadelphia Inquirer,* August 26, 1880, and "Rotten Eggs and a Conspiracy," *Philadelphia Inquirer,* August 28, 1880. No article subsequently appeared regarding the outcome of the case.
74. "Probable Murder," *New York Herald,* November 11, 1881; "New York Letter," *Philadelphia Inquirer,* November 14, 1881; and "Ad Magraw's Customer: The Son of a Noted Turfman and His Criminal Exploits," *Lancaster Daily Intelligencer,* January 22, 1883.
75. "The Sale of Erdenheim."
76. *1880 United States Federal Census,* NARA microfilm publication T9, roll 1159, page 240C, National Archives and Records Administration, Ancestry.com. For more on the Welch family and William A. Welch in particular, see Denise V. Berg, *The Welch Resort: A Hidden Treasure of Boulder County* (Lyons, Colo.: Kumpfenberg Ventures, 2006).
77. "Death of Aristides Welch," *Philadelphia Inquirer,* April 10, 1890. Other sources at the time of the transaction reported the forfeiture payment at $5,000. See "The Purchase of Erdenheim by Commodore Kittson," *St. Paul Daily Globe,* May 29, 1882, 1.
78. "Death of Aristides Welch," *Philadelphia Inquirer,* April 10, 1890.
79. "Death of Aristides Welch," *Turf, Field, and Farm,* April 18, 1890.
80. "Sporting Record," *New Hampshire Patriot and State Gazette,* April 14, 1881, 4.
81. "Death of Aristides Welch," *Turf, Field, and Farm,* April 18, 1890.
82. "The Gentleman Farmer as an Advertiser," *Printer's Ink,* 61:8, November 20, 1907, 17–18. Wilkes had become estranged from one of his two adopted children, George Wilkes, whom he cut out of his will shortly before his death, apparently under influence from George's sister Alicia. See "George Wilkes's Will," *New York Times,* October 10, 1885.
83. For another example of an advertisement of St. Jacobs Oil, see Ann Anderson, *Snake Oil, Hustlers and Hambones* (McFarland, 2004), 45–46. William Murrell, a correspondent to the *British Medical Journal,* analyzed the contents of St. Jacobs Oil for the publication. See Wiliam Murrell, "St. Jacob's Oil," *British Medical Journal,* May 5, 1894, 995.
84. "Aristides Welch: The Breeder of Iroquois and Parole," *Sacramento Daily Record Union,* December 13, 1881, reprinted from *New York Sportsman.* The original article—and its date of origin—could not be located in the *New York Sportsman.* The earliest known version of a brief

article listing Welch's endorsement occurred in the *Cleveland Plain Dealer,* October 15, 1881, suggesting that the *New York Sportsman* article must have appeared before this.

85. For an example of the brief, one-paragraph version of the Welch endorsement, see brief untitled item, *Daily Inter Ocean,* November 17, 1881. As evidence of how widely spread—and ultimately distorted—Welch's "endorsement" became, see untitled blurb, *Reno Evening Gazette,* November 21, 1881, and "What Was Seen at Erdenheim," *Spirit of the Times,* June 3, 1882, 501.
86. Untitled column, *Spirit of the Times,* June 11, 1881, 500.
87. Untitled column, *Spirit of the Times,* November 3, 1888, 560.

CHAPTER FOUR

From Horse Farm to Girls' School

Described by the press as showing "remarkable enthusiasm for a man no longer young," Commodore Norman W. Kittson was already well known among turf enthusiasts and much of the rest of the country by the time he purchased Erdenheim in 1882, but he was certainly not yet regarded as a major player on the turf. A resident of St. Paul, Minnesota, Kittson was considered new to the business, and his addition of "25 fine mares, imported and native" from his Midway Stud Park in Minnesota to the Erdenheim holdings seemed to bode well for his future. Others noted, however, that Kittson's investment seemed to be something of a lark, given that he "has not heretofore been conspicuous in breeding or turf events."[1] Indeed, when asked directly about it, Kittson seemed to confirm the whimsy with which he entered the field: "I can say that no consideration of gain led to my purchase of horses. I have worked pretty hard all my life and a man of my age needs amusement. I don't believe there is much money in turf enterprises, and don't expect to make a dollar. I want the horses for my personal pleasure and to introduce an improved breed into my State."[2] For these reasons alone, Kittson's purchase of Erdenheim was certainly unexpected, to say the least, among the horsey set. Add to these that Kittson had already sunk a substantial amount of money into racing interests in Minnesota, and his purchase seemed nonsensical. As one Minnesota newspaper noted at word of the sale, most in the racing world assumed that Kittson was done purchasing anything for a while after spending upwards of $250,000—a staggering sum for the day—on stock and the St. Paul Driving Park in Minnesota.[3]

Kittson's career prior to his entry into the horse breeding and racing business was a remarkable one, and today his name remains well known throughout Minnesota for his nonracing accomplishments. Born on March 5, 1814, in Canada, Norman Wolfred Kittson was the son of George and Nancy (Tucker) Kittson. He never completed more than a grammar school education, and by the age of sixteen, Kittson was serving at the direction of John Jacob Astor as an apprentice in the American Fur Company in the territories that would one day become Wisconsin, Minnesota, and Iowa. In 1843 Henry Hastings Sibley, the chief agent for American Fur, promoted Kittson to manager, assuring him an equal share of all profits Kittson secured in upper Minnesota. Buoyed by the promotion, Kittson engaged in an all-out trade war with the Hudson's Bay Company.

FACING: "New" Erdenheim and vineyard at close of day. Photo by Eric Plaag, May 2012.

His connections to the Métis people of the Red River Settlement (his wife, Élise Marion, was a Métis) provided him with a cooperative labor force and effectively allowed him to secure the northern boundary of the territory against Canadian intrusion, even if the profits did not make him as wealthy as he might have wished. He did, however, purchase real estate at St. Paul, and after serving in the legislature for Minnesota Territory from 1852 to 1855, he relocated to St. Paul, where he was elected mayor as a Democrat in 1858. Two years later the Hudson's Bay Company set aside any hard feelings about its decade-long competition with Kittson and hired him to be their purchasing and forwarding agent for Minnesota. With the introduction of steamboat service the following year, Kittson made a fortune overseeing both overland and river traffic for Hudson's Bay and various other private investors. When James J. Hill formed a competitive steamboat firm in 1871, Kittson persuaded Hill to join forces with him rather than destroying one another through competition, thus creating the Red River Transportation Company and earning Kittson the unofficial moniker of "Commodore." The introduction of rail lines to the region, and Kittson's early investment in them, only added to Kittson's wealth, culminating in his joint ownership of the St. Paul, Minneapolis, and Manitoba Railway by 1874.[4]

Kittson and Élise Marion raised at least six sons and two daughters, although some nonprofessional genealogical sources suggest he was married three times—first to the Swiss-born Sophie Parry or Perret (1817–89), with whom he may have had four children; then to Élise Marion (1831–68), with whom he may have had as many as nine children; and finally to Mary Cochrane (1842–86),

Commodore Norman W. Kittson, photograph by Zimmerman, circa 1880, Minnesota Historical Society.

by whom he may have fathered another six. In terms of the history of Erdenheim, the most important of these children were Louis and James E. Kittson. Born in St. Paul on April 2, 1855, Louis was sent at age four to Winnipeg to be educated by the French Jesuits at St. Boniface. In many ways following in the footsteps of his father, Louis entered the employment of the Hudson's Bay Company as a clerk at the age of seventeen, remaining with the company for eleven years. So wide and arduous were his travels that he was renowned for having walked eleven hundred miles in a single winter on snowshoes and for allegedly having "never [seen] a white man for six years." His father's purchase of Erdenheim brought Louis Kittson to Whitemarsh in 1883, where he assisted with daily operations of the stud. In 1889 he married Caroline D. Gordon, daughter of Robert Gordon of Springfield.[5]

It was Commodore Kittson's investments and notoriety, though, that sustained Erdenheim's reputation and drew even more attention to its operations than Aristides Welch had. So popular was the elder Kittson among his fellow Minnesotans that Kittson County, a new jurisdiction in the region from which he had served as a legislator, was named in his honor in 1879. His numerous investments in the fur trade, transportation, and speculation, which he maintained after the purchase of Erdenheim, continued to make him even wealthier. Entering the world of racing only added to that wealth, a move that proved necessary as his health began to fail and his ability to participate directly in the operations of the rail line and shipping companies waned. By 1884, with his attention almost entirely on breeding and racing, he had spent more than $2,000,000 on horses and farms, and he was believed to be a millionaire several times over. At Erdenheim alone he had more than a hundred thoroughbreds stabled.[6]

Erdenheim was already considered to be the prize stock farm of the racing world, "the Mecca towards which turfmen from all parts of the country made annual pilgrimages when the sales of yearlings took place," but Kittson's expansions and investments only magnified its reputation. Its facilities were considered among the best in the country, sporting "250 acres of falling land, with rich soil, excellent drainage and water facilities, and probably the best appointments of any stock farm in the United States." The three thoroughbreds (Alarm, Reform, and Lyttleton) and fifty plus brood mares that came with the real estate sweetened the pot substantially. By all accounts, the purchase was a bargain, and many speculated that Kittson intended to transform his Midway property into a farm for trotters, while Erdenheim would be his principal base of operations for thoroughbred breeding.[7]

The other great horsemen of the day—and the ever-speculative press—initially may have had their doubts about his investments at Erdenheim, but Kittson soon made his presence known with the sale of thirty-nine Erdenheim brood mares and other stock for $13,460 in November 1882 at the American Horse Exchange.[8] Kittson's intention in making the sale was "to reduce the stud within reasonable limits," given that he was already combining his own holdings with those at Erdenheim and had no intention of becoming a public breeder.[9] Meanwhile, at Erdenheim, Kittson wasted no time in expanding the farm's holdings from the 250 acres purchased from Welch to a total of almost 400 acres by striking a deal with Lewis A. Lukens to purchase his larger parcel of 147 acres, sixty-nine perches on the west side of the Wissahickon. The deal was consummated on May 21, 1883, for the price of $36,857.81—more than twice what Lukens had paid for both of the Mather parcels just thirty years earlier.[10]

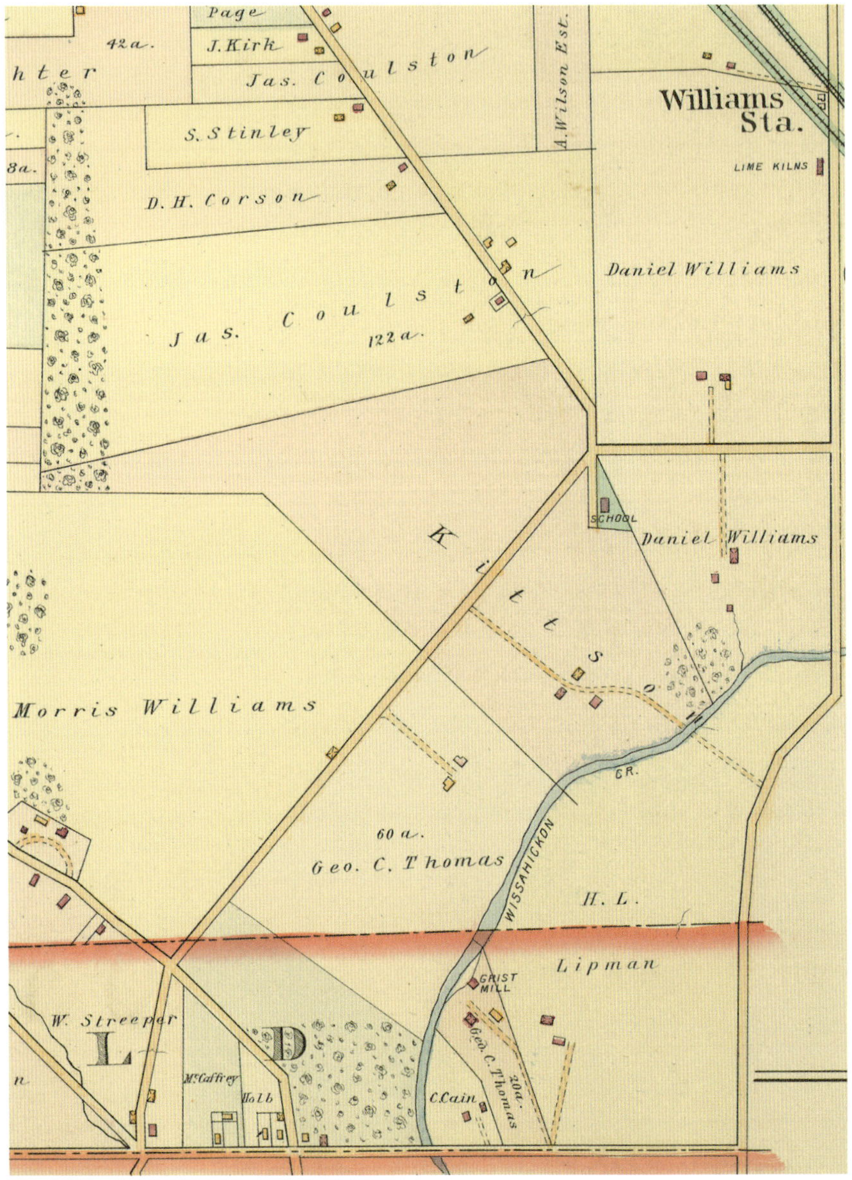
Page
42a.
J. Kirk
Jas. Coulston
S. Stinley
D. H. Corson
Jas. Coulston
122a.
A. Wilson Est.
Williams Sta.
LIME KILNS
Daniel Williams
Kittson
SCHOOL
Daniel Williams
Morris Williams
CR.
WISSAHICKON
60a.
Geo. C. Thomas
H. L.
Lipman
GRIST MILL
W. Streeper
L
D
McCaffrey
Holb
C. Cain
Geo. C. Thomas
20a.

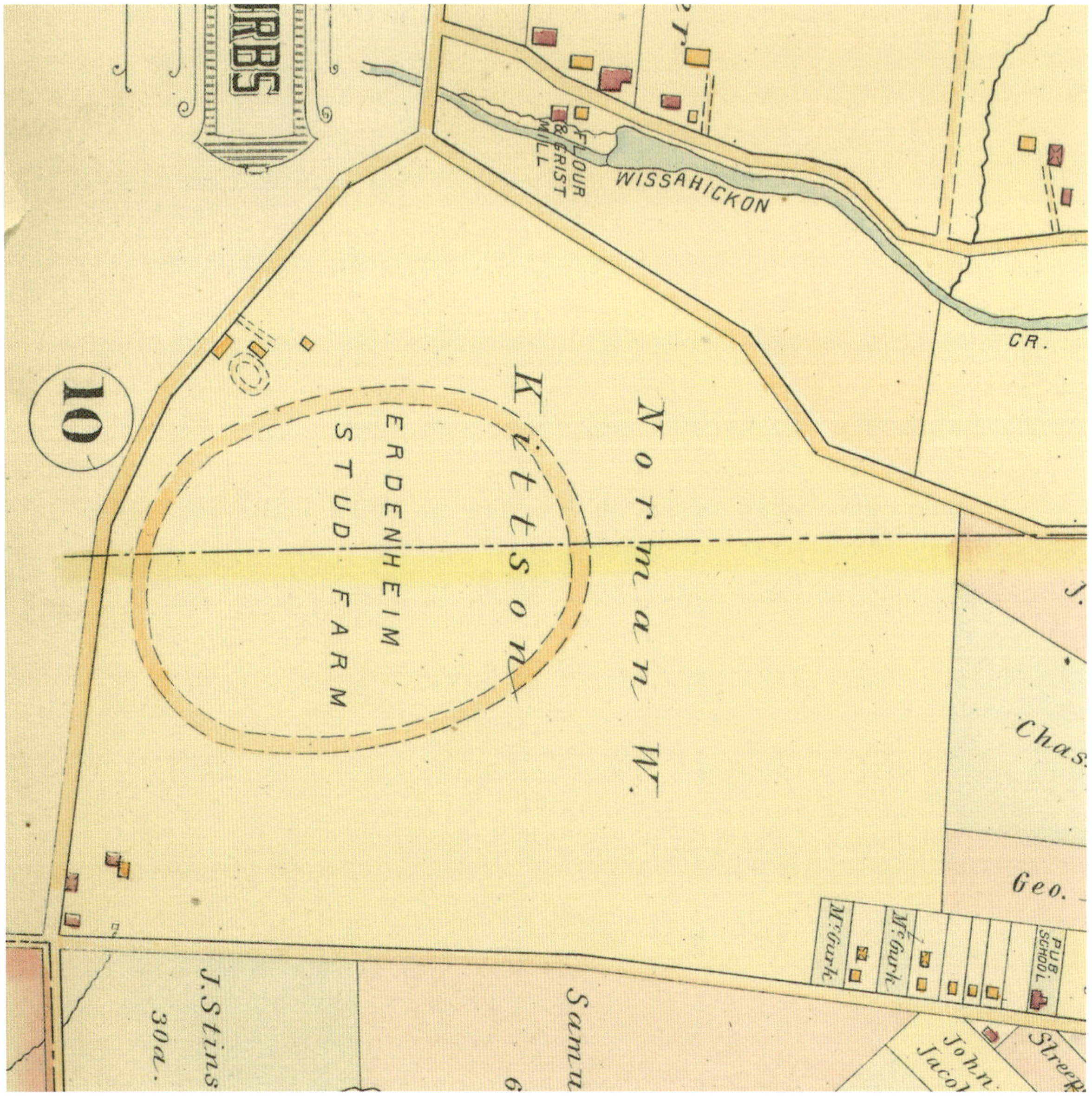
10
FLOUR & GRIST MILL
WISSAHICKON
CR.
Norman W. Kittson
ERDENHEIM STUD FARM
McGurk
McGurk
PUB. SCHOOL
30a.

Kittson's operations at Erdenheim were under the direct supervision of Major Josiah R. Hubbard, a lawyer from Tennessee who had been an officer with the Confederate army. Following the war he took up racing as a hobby, even as he continued to practice law in Nashville. Soon he was so immersed in the world of racing that he was hired in 1875 to author a series of letters to the *New York Sportsman* under the nom de plume of "Albion." Two years later he accepted the post of racing editor for the *Spirit of the Times* and moved to New York. In 1881 he retired from that paper and took up the role of agent on behalf of Commodore Kittson. It was Hubbard, in fact, who negotiated the sale of Erdenheim on Kittson's behalf. Kittson's other employees at Erdenheim included trainers Piercy Benn, William Midgely, James Lee, and Walter Rollins, a jockey named "Fitzpatrick," and John McCloskey, Welch's loyal stud groom for so many years. Kittson also employed an exercise boy and groom for every two horses on the farm. Supervision of the tillage of the remainder of Erdenheim fell to Abraham Creger, who had already been employed at Erdenheim for twenty years when Kittson took over the property.[11]

FACING: Maps showing the ownership of the various Erdenheim parcels as of 1893, as well as Kittson's reconfiguration of the Equestrian Tract with its new one-mile oval. Note that the "H. L. Lipman" indication at "Old" Erdenheim is incorrect, as explained elsewhere in this chapter. The maps also omit any detail of "Old" Erdenheim's buildings on the west side of Stenton. Detail from plate 10 (left) and plate 11 of *Atlas of the Properties in the North-West Suburbs of Philadelphia,* Philadelphia: G. Wm. Baist, 1893, Map Collection, Free Library of Philadelphia.

Kittson Transforms Erdenheim

Among Kittson's improvements at Erdenheim was a new one-mile track on the east side of Stenton Avenue, described by some observers as "one of the best in the country and as level as a billiard table." Built by William Mackey of Germantown for $17,000, the new track and an adjacent circular shed with horse stalls, living quarters for jockeys and stable hands, and an indoor track were something of an engineering marvel:

> The roof covers forty-five box stalls, a one-eighth mile track, a room for the trainer, apartments for the exercise boys and grooms, and a room for horse clothing, medicines, etc., leaving ample space for feed and straw on the second story. The outer circle of the track is protected from storms by the building, and the inner circle by weather boarding extending from the roof half way to the ground. The track will be devoted exclusively to training in the winter. The ground which encircles it will be laid out in walks and plats and will be embellished with the choicest flowers. A fountain will be erected in the center. The water from this, together with the refuse from the roof and ground, will be carried off by pipes laid beneath the surface. The apartment for the grooms and exercise boys is divided into a bed room, bath room, and a reading and smoking room, all of which will be heated by a heater and supplied with gas and water.[12]

Kittson also improved and increased the water supply to the farm, which was raised from a well by an Eclipse windmill built by Samuel L. Cowden, the Eclipse agent from Norristown, and pumped into two 10,000-gallon tanks in one of the barns. Kittson also maintained four additional barns, a blacksmith shop, and the half-mile track originally built by Welch, which Kittson eventually planned to abandon. The new one-mile track, meanwhile, was heralded as a superior improvement and regarded as "the widest private track in this section of the country," an addition that suggested "it may not be long before Philadelphia will have its Epsom, in Erdenheim."[13]

Kittson also made improvements to the dwellings at "Old" Erdenheim, ensuring that the "ancient house" in which Hubbard quartered would have "bath rooms, closets, stationary wash stands, and the appliances for furnishing hot and cold water"—all quite modern innovations for the typical country home of the early 1880s. Kittson also intended that "all the buildings will be illuminated either by gas or electric light," in such a sophisticated manner that "not only the buildings, but the surroundings, will be illuminated, and at night will be as light as when the sun is crossing the meridian."[14] Once Kittson obtained the Lukens parcel, he also effected a means of more easily conveying supplies from one side of the Wissahickon to the other by constructing a bridge across the stream at the present bridge location between "New" Erdenheim and "Old" Erdenheim in 1883. While the bridge piers here appear to have originally been of stone, as they are at present, maps from the period clearly indicate that the span was an iron truss.[15]

While Kittson's fame stemmed primarily from his activities as a breeder and a race enthusiast, it would be a mistake to conclude that agricultural pursuits at Erdenheim during this period were not also important. In 1884, for example, Erdenheim was reported to have generated a corn yield of 110 bushels to the acre, and the farm was noted for the vast fields of grass growing on the premises.[16] No evidence has been found, however, to suggest that Kittson carried on the sheep and cattle breeding operations that some owners before him perfected at Erdenheim.

Kittson's Horses

In addition to Alarm and Reform, the two famous champions that Kittson inherited from Welch with his purchase of Erdenheim, several other horses helped solidify Kittson's reputation on the turf. His successful racers included Issaquena, who won Monmouth's Hopeful Stakes in 1883, and Panique, who won both the Withers and Belmont in 1884, whereupon Kittson sold him for $14,000. Rataplan was another champion, winning the Emporium, Iroquois Stakes, and Travers.[17] In some respects, though, it was the horses that Kittson inherited from Welch that still drew visitors and whose fame still overshadowed all of Kittson's efforts at transformation, expansion, and innovation.

A reporter from the *Spirit of the Times,* who visited Erdenheim in 1888, waxed poetic about the then twenty-year-old Alarm, admiring his features and the grace with which he had aged:

> He looks very handsome and full of health, a robust bay with a Roman nose, and the massive forehead of the Stockwells. He has not swayed in the back as stallions are apt to do at his age, and looks good for seven or eight years to come. A better tempered stallion never lived. He was so as we remember him in training, "a bit of a slug" they used to say he was, but when in action no horse was ever faster. . . . He has the same grand Roman head, the massive neck, the round barrel, as round as the barrel of a rifle, and great spread of hips and enormous quarters.[18]

FACING: Panique, from *Spirit of the Times,* June 7, 1884, Keeneland Library.

PANIQUE, WINNER OF THE WITHERS STAKES.

Henry Stull
Erdenheim
1884

Reform was still at Erdenheim, too, in 1888, and the *Spirit of the Times* reporter was delighted when "Reform received us with 'open arms,' that is to say, he reared like a circus performer, and after rolling his eyes and lunging at us went off on all fours." Noting that the best of Leamington's colts, excepting Longfellow, were "vicious," he explained that Sensation was a "kicker," Harold was a "devil," Iroquois a "testy fellow," and "Onondaga a perfect man-eater." Nevertheless the visitor acknowledged Kittson's new blood on the farm by marveling at Pardee and noting that "we do not know when we have seen a more blood-like young horse," even though a case of thrush had forced McCloskey to fit Pardee with a special plate that slipped midrace and cost him the Lorillard Stakes.[19] Even Maggie B. B. was still there, a twenty-one-year-old mare that the reporter guessed could pass for an eight year old. "She carries heavily her pledge of Woodlands' affection," the reporter noted, "and she seems good for ten years yet, this mother of Iroquois, Harold, Panique, Pera, Francesca, and grandmother of Belvidere and Sir Dixon."[20]

Among Kittson's other champions were Johnston and Little Brown Jug, described in early 1884 as "the two fastest pacers the turf has ever known," each holding records of 2:10 and 2:11.75, respectively, the fastest miles ever run to that point. Kittson also possessed a large collection of trotters, including Fanny Witherspoon, So So, Minnie R., Von Arnim, and Pilgrim, all driven by John Splan, a man in his early thirties believed by many to be, "in point of ability as a driver and trainer . . . second to no man in the country, although there are three or four that are his equals."[21]

NORMAN W. KITTSON

FACING: Alarm, by Henry Stull, from *Spirit of the Times,* February 16, 1884, Keeneland Library.

Racing silks worn by Norman W. Kittson's jockeys, based on various newspaper accounts and Frederick M. Burlew's *American Racing Colors: 1865 to 1915,* volume 1, Keeneland Library. Design by Kevin Archie.

By late 1884, though, there were hints that not all was right at Kittson's Erdenheim. The performances from his runners and trotters had been disappointing for the better part of a year, and several of the horses he had paid vast sums for had failed to meet expectations.[22] A visiting reporter that fall noted that Alarm was in excellent condition, but, "nevertheless, Commodore Kittson is so overstocked with stallions that he would let him go, should anyone fancy him enough to pay the price." Reform, meanwhile, was described as "a bundle of nerves," acting "like a peacock when McClosky [*sic*] led him out. He turns in the left forefoot, owing to a cut by treading on a piece of glass." While many of the other horses were described as being in good form, there was also a long list of other injuries among them, something that had rarely been an issue under Welch's ownership. Issaquena was struggling with a weak flexor tendon, while Pardee suffered from a "shriveled fore foot," an injury described as being worse than it should be because it was not properly treated from the beginning.[23]

Perhaps the most devastating news, though, came on September 30, 1884, when flames broke out in the upper windows of a two-story barn while a large group of horse owners were exercising their horses at the nearby track across Stenton Avenue. Patrick Slavin, the head farmer, discovered the smoke pouring from the windows and summoned Superintendent Hubbard, who rang the bell for assistance. Frantic telegrams to Norristown and Philadelphia calling for fire suppression equipment went unheeded, the authorities there insisting that they were not permitted to leave the city limits. Alarm and Reform were hustled to the safety of another building, while the other horses were moved to the center of the one-mile oval. From there farmhands and horse owners alike joined in a bucket brigade to stop the spread of the flames. In spite of their efforts, Ailene, a $1,000 three-year-old filly, perished in the fire.[24]

The barn—described as "about 35 by 100 feet, with an L-shaped extension 20 by 90, used as a hay house"—and an adjacent "two-story building, 28 by 16 feet, the ground floor of which was used as a stable and the second floor . . . filled with hay" were completely destroyed. Also lost to the flames were seventy-five tons of hay, eighty bushels of wheat, and thirty tons of straw. Louis Kittson soon arrived at the scene, insisting that the entire loss of approximately $30,000 would be covered by insurance. Slavin, the head farmer, was at a loss to explain the cause of the fire, noting that no fires or lanterns were kept in the barn, and smoking was forbidden in all the farm buildings. Nevertheless the fire was not believed to be the result of arson.[25] As with any large farm housing hundreds of thousands of dollars' worth of investment, the risk and fear of fire at Erdenheim must have always been lurking in the minds of its owners, but the circumstances of this fire seem curious. As one reporter remarked after the fire, noting that a new stable would be built once the burned walls of the old one were brought down, "The stables were too much bunched at Erdenheim, and the wonder is that the fire, having got under strong headway, did not wipe them all out."[26]

The 1884 fire, however, was not the end of Kittson's troubles at Erdenheim. On the morning of October 15, 1885, watchman Peter Daran discovered flames licking from the roof of the brood mare stable, about five hundred yards south of the main stable area. The old stone barn, which may have been the 1870 stone barn built by Welch on the east side of Stenton, contained farm implements and a vast store of hay, all of which were totally destroyed, leaving only "the smoke-blackened walls of the stone barn." Fortunately none of the animals were in the barn at the time of the fire. Erdenheim's faithful breeder, John McCloskey, had allowed the entire stable of mares to run in the paddock under the bright moonlight. While frightened by the fire, none of them were injured. This time the cause was attributed to "the carelessness of tramps, who sometimes sleep in the stable loft."[27]

Thomas Jeremiah Healey

Among Commodore Kittson's employees at Erdenheim in 1885 was a young stable hand named "Strawberry Tom" Healey. Born on July 16, 1866, at Fordham, New York, Healey's stay at Erdenheim was relatively brief. He had knocked around since the age of fifteen in stable work at several farms in other parts of the country, but his nineteenth birthday found him at Erdenheim. All indications are that Kittson's retirement from the turf the following year all but ended Tom Healey's stay in Whitemarsh. Nevertheless he pulled up his boot straps, found work elsewhere, and went on to become one of the legends of American horseracing.

By 1888 Healey was in the employ of Walter Gratz of Philadelphia, where he worked as a trainer for horses that made an impressive showing that year at various races. Within just a few years, Healey was in charge of one of the largest public stables in America, managing horses belonging to dozens of owners from throughout the country. In 1896 he found work with the president of the Saratoga Association, for whom he remained employed as a trainer for the next thirty-five years. As ownership of the association changed over the next eight years, Healey remained at his post, training the horses of various owners who stabled there until 1937, when the association dispersed. Four years later the Laurel Park course in Maryland hired Healey as their steward. As one contemporary

Thomas J. Healey, left, with L. A. Beard, 1933. Cook Collection, 1565, Keeneland Library.

put it, "Despite his advanced age, horsemen were as nearly unanimous as horsemen can be in lauding him as one of the country's best stewards."

Over the course of his career from 1907 to 1936, horses under Healey's care brought in a total of more than $2.8 million for their owners. His winners included Pillory (Preakness and Belmont Stakes), Top Flight, Equipoise, The Parader, Campfire, Olambala, and the Preakness winners Vigil, Display, and Dr. Freeland. For a time he even owned and raced his own horses, including stakes winners Lion D'Or and Tatterdemalion. Healey died on October 7, 1944.[28]

Kittson Retires from the Turf

In March 1886 Kittson announced his intention to retire from trotting matches, a move that was not met with particular surprise, given that Kittson had not been active in that class of racing for at least a year. He assured the racing community, though, that the decision would not affect the future of his horses at Erdenheim.[29] Several months later, however, Kittson announced his intention to withdraw from the turf altogether, as he was "at an advanced age, and feels indisposed to continue an active turf career with its incidental cares and anxieties." Nevertheless, he insisted, Erdenheim would continue under his guidance to be a top-notch breeding operation, and he promised through the press that "the old Chestnut Hill stud will revive its old glories of the Leamington era."[30] Kittson's long-time manager, Major Hubbard, remained at Erdenheim, directing his energies not toward the racing that had dominated his time for the previous four seasons but instead toward breeding. Just six months later, though, Kittson stunned the racing world when he announced his plan to sell his entire stable of yearlings and race horses, consisting of thirty-six animals that brought a total of nearly $30,000 at auction. Hubbard appears to have called it quits at this point, too; he spent his remaining years in retirement as well, before he died in September 1889.[31]

Commodore Kittson's death, while perhaps not unexpected for a man of his age, nevertheless came as a shock to the racing world. While traveling on the Omaha train northbound for St. Paul on the night of May 10, 1888, Kittson entered the dining car, seated himself, then died suddenly just outside Chicago.[32] As for Kittson the man, the occasion of his death allowed the papers to recount his reputation. Some said he was renowned for "never having wronged an Indian in any transaction, and many of the tribes would sell their furs to no one else." He was allegedly fluent in their native tongue, and was said to possess "the Indian trait of taciturnity in a great degree . . . and rarely spoke of his business or personal affairs, and disliked to talk about the affairs of others." Just days before his death, he advertised his remaining horse stock at auction, effectively announcing his intention to get out of the breeding business as well, although at the time of his death he was still serving as president of the Minnesota Trotting-Horse Breeders' Association and the Northwestern Breeders' Association.[33]

After Kittson died, responsibility for divesting the farm of some of its stock fell to the estate's executor, the St. Paul Trust Company. As a result, six months later, on a cold, rainy day with an unexpectedly small crowd on hand, St. Paul sold off much of Kittson's stock in an auction at Erdenheim, the sale totaling fifty-two animals—fifty of them horses, six of those stallions—and bringing in just under $50,000 in revenue. The farm was also put up for private sale at an asking price of $300,000. Among the horses sold were nineteen-year-old Alarm

FACING: Advertisement regarding the sale of the Erdenheim Stud, *Spirit of the Times,* November 3, 1888, Keeneland Library.

THE ERDENHEIM STUD AT AUCTION.

By Order of the St. Paul Trust Company, Executors of the Estate of N. W. KITTSON, Deceased, I Will Sell at Public Auction, Without Reserve, the

ERDENHEIM BREEDING STUD,

CONSISTING OF SIX STALLIONS AND FORTY-SEVEN BROODMARES.

ALARM, sire of Panique, Himyar, Gabriel, Breeze, Startle, Soubrette, Pardee, Danger, etc.
REFORM, sire of Rataplan, Miss Lumley, Brown Duke, Ittilala, Florence E., Little Fred, Calera, Radiant, Tessa K., etc.
DALNACARDOCH (imported), sire of Drumstick, Carnegie, Pat Donovan, etc.
WOODLANDS (imported), sire of Brynwood, Suitor, Woodson, Merci, etc.
RATAPLAN, winner of the Emporium, Travers and Iroquois Stakes, Coney Island Cup, etc.
PARDEE, winner of the Tidal Stakes. A very speedy race-horse and finely bred.

FORTY-SEVEN BROODMARES, including

MAGGIE B. B., dam of Iroquois, Harold, Panique, etc.
WALTZ, dam of Glidelia, Hop, Racket, etc.
LADY LUMLEY (imp.), dam of Rataplan, Miss Lumley, Sayre, Luminous, etc.
MEGARA, dam of Spinaway, La Belle N., etc.
YORKSHIRE LASS (imp.), dam of Little Fred, Rustler, Burr Light, etc.
RACHEL, dam of Refrain, Radiant, etc.
ALGEBRA (imp.), dam of Donald A., Woodson, etc.
ESSAYEZ II. (imp.), dam of Issaquena, Ittilala, etc.
SISTER OF MERCY, dam of Pardee, Heck, Merci, Tessa K., etc.
BLUE LODGE, dam of Aura, etc.
LADY SALYERS, dam of St. Paul.
AUSTRALIND, dam of Circassian, Austrianna, Hayward, etc.
SYRIA, dam of Brynwood, etc.
VICTORIA (imp.), dam of Regulus.
TEMPTATION, dam of Rivet.
CLARA (imp.), dam of Cartoon.
ZICKA, dam of Leather Stocking, Benedictine.

Together with the famous race-mares **GLIDELIA, ISSAQUENA, LUMINOUS, BABY, FAIRWATER ALBIA, ARTIFICE,** etc., from distinguished racing a producing families.

THE SALE WILL BE HELD AT

ERDENHEIM, CHESTNUT HILL, PHILADELPHIA, PA., THURSDAY, NOV. 8, 1888, COMMENCING AT 1 P. M.

A special train will be run from New York City on day of sale, of which due notice will be given.
Catalogues will be ready Sept. 1, and can be had on application to L. KITTSON, Chestnut Hill, Philadelphia, Pa., or to

S. D. BRUCE, Auctioneer, P. O. Box 362, New York City.

(who went for $2,050) and seventeen-year-old Reform, who went for a mere $350. Twenty-two-year-old Glidelia brought the highest price of $4,100. Maggie B. B., the famous dam of so many champions from Erdenheim, brought a paltry $1,100, but the buyer permitted the Kittsons to keep her "for sentimental reasons."[34] In May 1889 the American Exchange in New York auctioned off thirty-one Erdenheim yearlings, securing approximately $18,000 for the estate. Prices ranged from $150 for Vanita, a chestnut filly by Reform, to $3,600 for Sister of Mercy, a bay filly by Alarm.[35] The following year Louis and James E. Kittson, Commodore Kittson's heirs to Erdenheim, sold off another small lot of Erdenheim's yearlings.[36]

Dealing with the rest of the Kittson estate, however, was more complicated, in large part because Jesse P. Farley of St. Paul, Minnesota, had brought suit against Commodore Kittson and his partners many years earlier, alleging that Commodore Kittson had brought him in as a one-fifth share partner on his purchase of the St. Paul, Minneapolis, and Manitoba Railroad in 1880. After the U.S. Supreme Court heard a plea of abatement in the case in 1887 and returned it to the Circuit Court of St. Paul for resolution, the trial was not heard until September 1889. In light of the massive value of Kittson's estate by that time, Farley was seeking upward of $4,000,000 in damages, which just happened to be the approximate value of Kittson's estate. The case was finally settled in the defendants' favor in September 1889.[37]

Settlement of that case, of course, meant that the rest of the Kittson estate could be settled, too. That said, Kittson's will stipulated that one half of the land at Erdenheim should go to Louis Kittson and the other half to his brother James,

Louis Kittson, from Ruoff's *Biographical and Portrait Cyclopedia of Montgomery County*, 1895.

and it appears from the deed record that the division of Erdenheim was accomplished in just this fashion.[38] Newspaper reports, however, indicated that the entire farm was put up for auction in three separate tracts. Louis Kittson purchased "the main estate, 'Erdenheim,' consisting of 247 acres and suitable buildings," for $300 per acre, apparently doing so through a surrogate named James G. Weldon for the sum of $74,000. Robert N. Carson, the former president of the People's Passenger Railway Company, purchased the other two tracts—one containing 57 acres and "a large, old-fashioned mansion," the other comprising 91 acres "suitable for building sites," these together being the old Lukens lands—at $400 per acre for the first plot and $300 per acre for the latter. A fourth tract of 3 acres (apparently part of the original Erdenheim parcel) also allegedly went to Carson, although no deed for this transaction appears to exist. Louis Kittson, in turn, announced his intention to transfer control of his portion of the former estate to a superintendent who would operate the revived stock farm on his behalf.[39] Further complicating matters was the fact that James E. Kittson transferred his one-half interest in the Erdenheim parcel to Louis Kittson on November 24, 1893.[40]

Even as the disposition of the Erdenheim lands and the settlement of the Kittson estate among his eleven heirs came to a close, a new challenge emerged in January 1894. That month a Native American woman from Wisconsin, claiming to be Margaret Robinson Kittson and thus yet another of Kittson's various brides, demanded a third of the estate. She also claimed that her two children, Sarah and David Kittson, were Commodore Kittson's offspring, and she produced what she alleged was a marriage certificate from Little Chute, Wisconsin, dated May 15, 1833. Among the various problems with her case, though, were the fact that Little Chute did not exist in 1833, Wisconsin was not yet a territory then, and the member of the clergy whom she claimed had signed the certificate was living in Ohio at the time of the certificate's alleged date. The case finally came to trial in March 1896, at which time the trial judge determined that no legal marriage ever existed between Robinson and Kittson, and the probate court was permitted to proceed with its final certification of the division of the estate.[41] Kittson's only public bequest from his will, meanwhile, was a sum of $10,000 to the Masonic Building Association. Everything else went to his children or other family members.[42]

By then Louis Kittson appears to have grown weary of maintaining his operations at Erdenheim. Indeed there were signs immediately following Commodore Kittson's death that Louis and James Kittson had other plans for Erdenheim, given its estimated value of $500,000 for the land alone. As early as 1888, breeding operations at Erdenheim were leased out to Edward Kelly, but even this outsourcing of Erdenheim's principal industry wasn't sufficient to keep the interest of the Kittsons in maintaining the farm. In December 1891—after most of Erdenheim's stock had been sold—blurbs began surfacing in the papers suggesting that "Erdenheim . . . is too valuable to be used for breeding purposes. . . . It will soon be cut up into plots suitable for suburban homes."[43] Fortunately nothing came of this speculation, but it remained clear that Louis Kittson had no intention of hanging on to Erdenheim for the long haul. As early as April 1896, Kittson struck a bargain with Robert N. Carson to purchase the remainder of the Kittson holdings at Erdenheim. On June 12, 1896, Louis Kittson finally parted with what remained of Erdenheim in a transaction involving four distinctly described parcels totaling nearly 247 acres for the sum of $32,500.[44]

Life after Erdenheim for the Kittson brothers was a mixed bag. James E. Kittson developed a morphine and cocaine habit toward the end of the century, one that apparently incapacitated him from being able to function properly for several years. Toward the end of 1900, James told his brother Louis that he

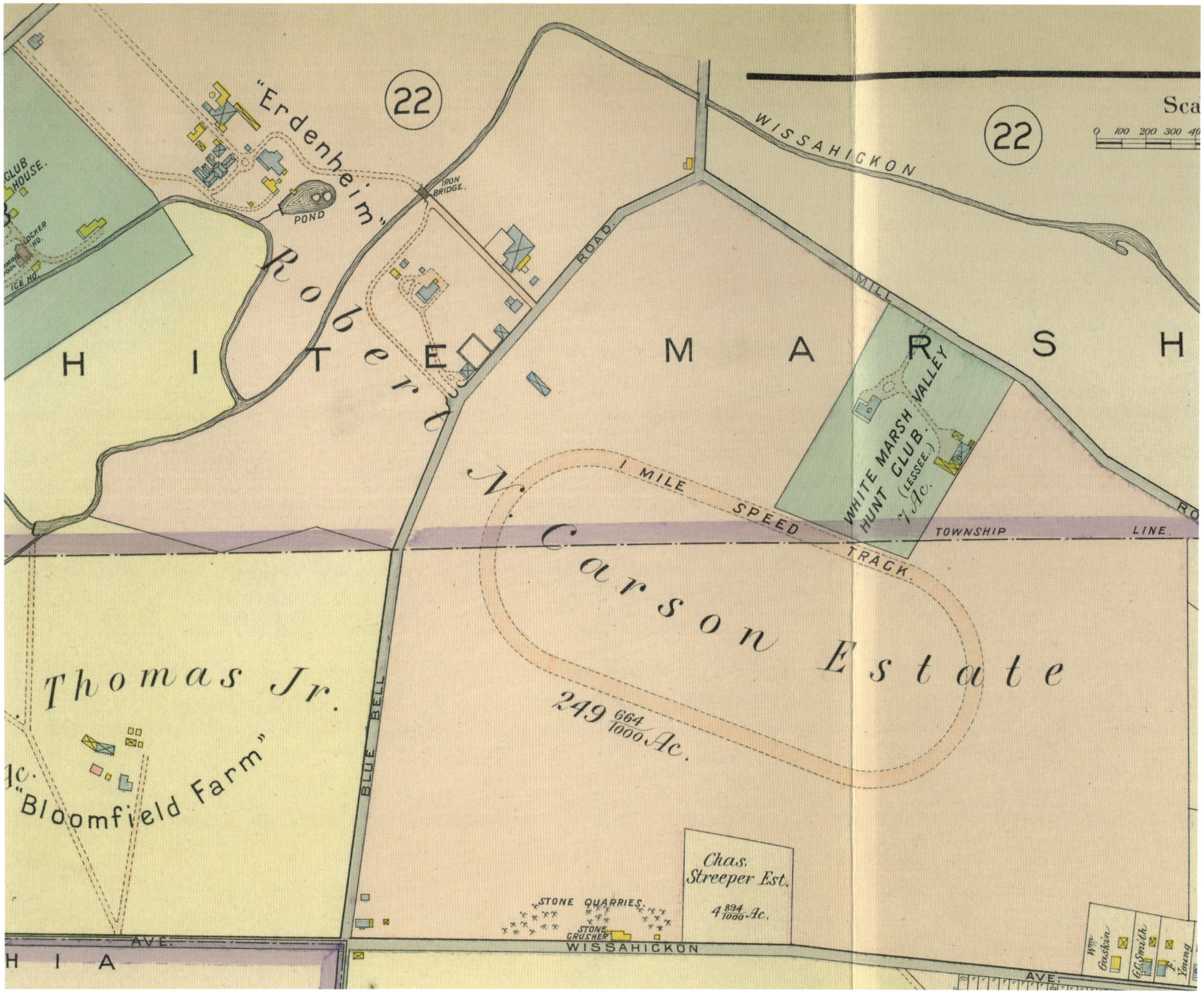
"Erdenheim"
22
22
Sca
0 100 200 300 400
CLUB HOUSE.
POND
IRON BRIDGE.
ICE HO.
WISSAHICKON
ROAD.
MILL
H I T E M A R S H
Robert N. Carson Estate
249 664/1000 Ac.
1 MILE SPEED TRACK.
WHITE MARSH VALLEY HUNT CLUB. (LESSEE.)
TOWNSHIP LINE.
Thomas Jr.
"Bloomfield Farm"
BLUE BELL
Chas. Streeper Est.
4 894/1000 Ac.
STONE QUARRIES.
STONE CRUSHER
WISSAHICKON
AVE.
H I A
Wm. Gaskin
G.G. Smith
Young

Robert Nethermark Carson, circa 1900, courtesy of Carson Valley Children's Aid.

FACING: The "New" and "Old" Erdenheim parcels as they appeared in 1909. Note that "Iron Bridge" indicated at the crossing of the *Wissahickon,* as well as the fact that the "Erdenheim" name has shifted from the historic location to the old Lukens farm. Detail from *Atlas of Properties on the Reading Railway, Embracing Cheltenham, Abington, Springfield, and Parts of Moreland and Whitemarsh Townships,* Philadelphia: A. H. Mueller, 1909, plate 20, Map Collection, Free Library of Philadelphia.

planned to break off his usage cold turkey in an apparent bid to pull his act together. On November 16, 1900, however, Louis Kittson and his wife awoke to the sound of a heavy thud from the room James kept at their home in nearby Fort Washington. James had shot himself in the head with a shotgun at the age of thirty-nine, making himself the third suicide among Erdenheim's previous owners in a twenty-year period.[45]

Of Louis Kittson, the historical record tells us very little after 1900. There are notices regarding social events in which he participated, and some indication that Louis and his surviving brothers parlayed their father's shares of the *St. Paul Daily Globe* into full ownership of the paper by 1902.[46] After that date, however, Louis Kittson's name is nearly absent from the Minnesota and Pennsylvania papers. The last clue to how he spent his final days surfaced in 1920, when it was announced that he had been disqualified from his position as acting postmaster at Fort Washington on account of his age, a post he had been fighting to keep for at least seven years.[47] How he spent his remaining years, as well as the date of his death, are not known.

Robert Carson's Erdenheim

To this day Robert Nethermark Carson remains something of an enigma, and by most accounts, he seemed to be as much to his contemporaries when he came to Erdenheim. Indeed it boggles the mind that so little documentation could survive on such a prominent Philadelphia figure from the late nineteenth century. Born in March 1844 at Philadelphia, Carson made his early career working for the banking firm William H. Shelmerdine and Company, which invested heavily during that period in traction (or streetcar) company stocks. Within a relatively brief period of time after Carson left the firm in 1881, he possessed a controlling interest in various local rail systems linked to the Philadelphia Rapid Transit Company, and in short order he was also serving as the president of the People's Passenger Railway Company, a position he maintained until his retirement in 1893. Carson also served as the proprietor of the Hamilton Apartment House at 1334 Walnut Street in Philadelphia, where he lived for many years.[48] In the years after he left People's Passenger, he organized a new company devoted to electrically powered rail lines in the Philadelphia suburbs—the Interstate Trolley Company. Between his earlier investments in rail travel and his prominent position at the Interstate Trolley Company, we can only guess that the traction cars made him rich. Carsonia Park, an amusement attraction located near Reading, was named in his honor.[49]

Of Carson's private life, we know almost nothing. The 1900 Census lists Carson and his wife Francis Isabel Flickenger (Carson), who was born in May 1846, as residing at "New" Erdenheim and indicates that they had been married for five years. They shared their home with an eleven-year-old niece, Retta Flickenger, and three female servants. The Carsons, we know from the disposition of his estate, never had any children of their own.[50] The Carsons summered

FACING: "Old" Erdenheim as it appeared circa 1895. The children in the left foreground are not identified. Springfield Township Historical Society.

BELOW: The old stone barn/stable, built by Welch in 1870 on the east side of Stenton Avenue, as it appeared in 1907. From *Philadelphia Record,* October 27, 1907, courtesy of Springfield Township Historical Society.

frequently at Newport, Rhode Island, especially after his purchase of Erdenheim, and he was known to have a "large white steam yacht," the *Marietta,* that he used for his frequent trips to Newport.[51] It is certainly possible that Carson, as someone frequently announced on the society pages of that town, may have come to know Atherton Blight, the former owner of Erdenheim who also spent his summers there during this same period, but no direct evidence survives to whet any curiosity about the two Erdenheim scions meeting and discussing the old farm.

At first, after Carson's purchase of the Erdenheim parcels, rumors circulated that Carson planned to turn the farm into an excursion ground for the Union Traction Company, but these were soon dismissed.[52] In many respects, in fact, the more things changed at Erdenheim, the more they stayed the same. John McCloskey, the trusted breeder and groomsman under both Aristides Welch and Commodore Kittson, was still employed there, now saddled with the title of "custodian" in addition to his many other responsibilities. He lived on the premises, too, sharing one of Erdenheim's several other houses with his sister, three sons, a servant, and four boarders. As for "Old" Erdenheim, the homestead in which George Hocker had run an inn and fended off an intruder, and in which all of Erdenheim's other owners from General Scheetz to Aristides Welch once lived, it too still stood watch over the Wissahickon. By 1900 it was home to R. Nelson Buckley, Carson's nephew and eventual estate executor; Buckley shared the residence with his mother and a young servant.[53] One commentator observed in 1896,

> It stands there, white and stately in the sunlight, overlooking the lovely Wissahickon and so suggestive of by-gone hospitality that it almost seems to speak and bid one enter. It is vacant now, but in excellent order, and it does not need an active imagination to fancy it filled with the fair women and brave men of the old days or to hear the voices and laughter of the jovial spirits who were prominent in the affairs of our turf in its earlier history. The style of the house is Colonial with great pillars running up to the roof, and inside the halls and rooms are large and comfortable with great fire-places and large windows.[54]

In 1896 Carson's "Old" Erdenheim parcel still retained Welch's old stable that he had built specifically for Leamington in 1876, now "fitted up as an office" and located within 100 feet of the old house, probably just above the Wissahickon. To the left of the house, closer to Stenton Avenue, there were now three horse graves in the yard, the most recent belonging to Maggie B. B., although she did not yet have her own granite slab to commemorate her. Just beyond this, adjacent to Stenton Avenue, was the private stable Welch had built for himself, "in excellent preservation and quite in keeping with the beauty of the old place." Immediately opposite "Old" Erdenheim, on the east side of Stenton Avenue, were

Etching of the private stables built by Welch and located on the west side of Stenton Avenue near the graves of Leamington and Flora Temple. From *St. Paul Globe,* August 24, 1896.

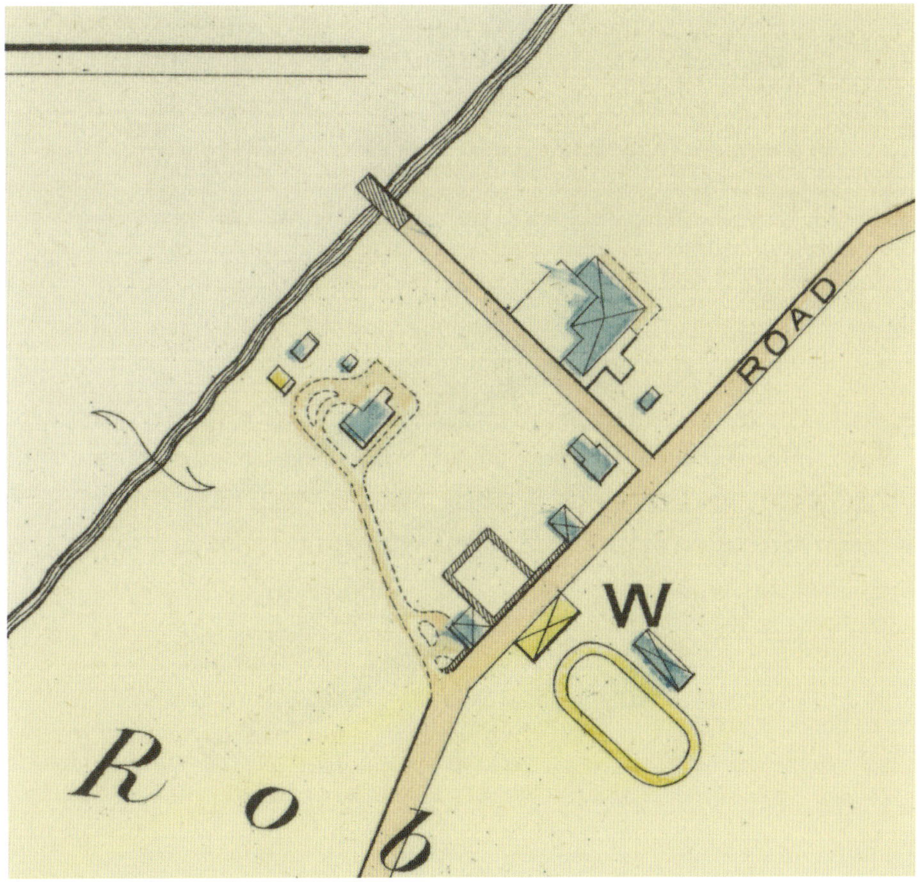

Detail of the Erdenheim buildings, circa 1896, from *Atlas of Cheltenham, Abington, and Springfield Townships and Vicinity, Montgomery County, Pa.*, plate 18, 1897, Map Collection, Free Library of Philadelphia. The blue building on the east side of Stenton Avenue is the 1870 Welch stable still standing at Erdenheim. Kittson's training track is the oval in yellow on the east side of Stenton; the yellow indicates frame construction, referencing the fact that the oval was covered. The yellow rectangle to the west of "Old" Erdenheim was probably the old Leamington stable, converted to an office in 1896.

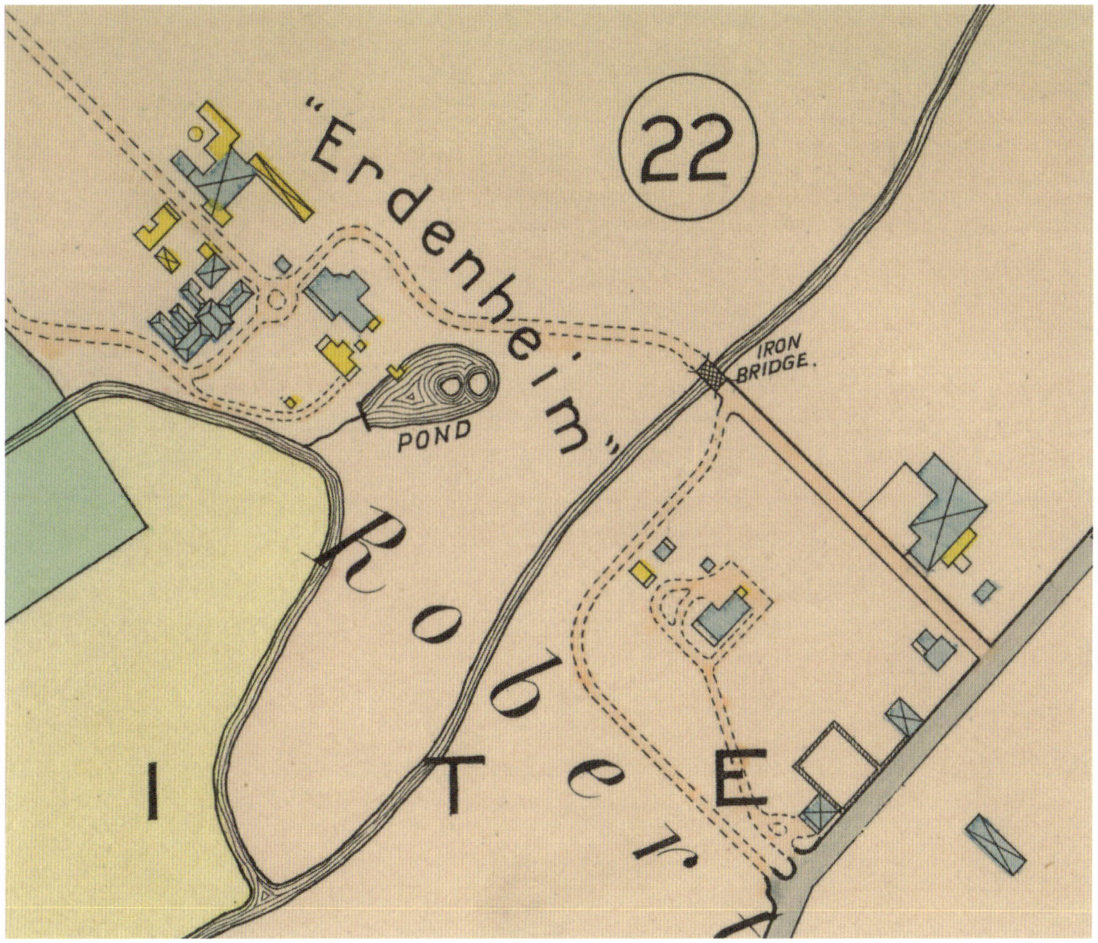

Building detail for "New" and "Old" Erdenheim, 1909, from Kiser and Lathrop's *Atlas of the Properties on the Reading Railway,* 1909, plate 20, Map Collection, Free Library of Philadelphia.

two stables originally built by Welch (including the 1870 stone stable that still stands at this location), plus the circular training stable with its small, interior, training oval. Under Carson's ownership these stables were used to house the horses of other significant owners, including Theodore Cuyler Patterson's Abdul Hamid II, Cupid, Omar, and Kasim.[55]

By 1890 Kittson's self-promotion had made the Erdenheim name so famed that it had not only been attached to a horse, but it also served as the appellation of one of the races—the Erdenheim Welter Handicap—at Morris Park (then in Westchester County, New York), home of the Belmont Stakes that year and the Preakness from that year through 1904.[56] The Erdenheim name would continue to be linked to the races themselves in the years to come, serving as the name of a different stakes—the Erdenheim Handicap—at Havre de Grace, Maryland, as early as 1915.[57]

But other racetracks were not the only place where the Erdenheim name was revered. A nearby community, located to the south of Erdenheim Farm and formerly known as Wheelpump for a tavern of that name, adopted the name Erdenheim in 1892. Under Carson's ownership Erdenheim's one-mile oval also became a destination for horse fanciers and the Philadelphia social set alike, who thrilled to observe the races from the new clubhouse Carson built adjacent to the track.[58] One account from 1906 trumpeted the trotting races scheduled for the following day, where Erdenheim's proprietor believed the track record would be broken. Steeplechase was scheduled as well, with "new and expensive hurdle jumps . . . making it very attractive and interesting to see the horses galloping across the fields at full speed, and at no time out of sight of the onlookers." The day promised a one-mile flat race and a high-jumping competition as well.[59] Such events were not new to Erdenheim; as early as 1904, races were being held there "under the auspices of the Erdenheim Trotting Association."[60] In 1909 the Erdenheim Trotting Association made further improvements to the oval, widening the stretches and building up the turns to make it "one of the best in the State today."[61]

The real showcase of Erdenheim, however, was "New" Erdenheim, standing on the opposite side of the Wissahickon from its original namesake and rising up on a high hill. Here Carson took the old Lukens home and remodeled it, and in the fields to the west of the house, he added a domed conservatory and several barns, thus completing the transformation of the property inaugurated under Commodore Kittson's ownership.[62] Images from circa 1903 indicate that the main house was three stories tall, the top floor replete with dormer windows on both the front and rear. A one-story porch ran along the front of the house, while a crenellated one-story addition (probably built by the Kittsons) stood on the west side of the house (outside the original west stone chimney stack). A gambrel-roofed, two-story addition with two interior stone chimneys and dormers on the second floor stood to the east of the original house (outside the original east stone chimney stack). Adjacent to the main house to the southeast there was by 1909 a substantial carriage house, and a little farther south of this, just above the oxbow of the Wissahickon, was a small tennis clubhouse. Carson also built a small "summer house"—in reality, a rusticated gazebo-type structure—along the north shore of the Wissahickon oxbow. Further west, though, just beyond the traffic circle in front of the old house, were the more significant additions, including a massive stable and greenhouse complex of stone, the aforementioned conservatory, and numerous outbuildings to support the needs of the farm and

FACING: "On the Wissahickon. Home of Robert Carson" (DAMS 468), circa 1903, Thomas H. Shoemaker Germantown and Philadelphia Portraits and Views Collection (V86), folio 4, Historical Society of Pennsylvania.

the house. Meanwhile, to the immediate east of "New" Erdenheim, Carson transformed the low area there (which was known to possess a spring) into a pond, dammed up the natural drainage of this field into the Wissahickon oxbow, and built a small boathouse on its periphery.[63]

By 1906 the farm at Erdenheim contained a total of 360 acres, about 110 of which were engaged in cultivation for feed or other produce, with the remaining land in permanent pasture. The barns that year contained over one hundred box stalls, and the various fields at Erdenheim were divided by high post and rail fences, which helped ensure the safety of Erdenheim's various animals. Morris Williams, meanwhile, was listed as being the proprietor in "control" of the farm. Hugh Joyce was also there, serving as trainer to a number of horses being stabled at Erdenheim, giving lessons in riding and jumping to visitors to the farm, and selling approximately twenty-five of his own horses that he kept stabled there.[64] Horse breeding operations, renewed there under the supervision

FACING: "On the Wissahickon. Rustic Summer House at Erdenheim, Built by Robert Carson" (DAMS 471), circa 1903, Thomas H. Shoemaker Germantown and Philadelphia Portraits and Views Collection (V86), folio 4, Historical Society of Pennsylvania. The greenhouse complex is visible in the distance at upper left, and the house at "New" Erdenheim can be seen through the trees at upper right.

"New" Erdenheim, as it appeared in 1901. A small footbridge over the lower end of the pond is visible to the left of the house, while the greenhouse and conservatory complex can be seen at the extreme right. From Moses King, *Philadelphia and Notable Philadelphians,* 1901, courtesy of the Free Library of Philadelphia.

of John McCloskey, were also once again making a name for the farm. By "interbreeding Welsh, Connemara, and Russian pony mares with thoroughbred and Arab blood," McCloskey was succeeding in creating a new line of stock that "abundantly prove[s] the correctness of the lines there pursued in breeding for foundation purposes and for type."[65] When McCloskey died circa 1904, his son John J. McCloskey took on some of the responsibilities for training while turning over control of operations completely to Morris Williams, who appears to have purchased the actual stock operation in 1903.[66] All told, there were seventy-five horses at Erdenheim in 1906, many of them boarded by other owners. In association with the breeding operations, such well-regarded horses as Prince Ralph and Josh were standing stud, while many other horses were there to train, including racers Cinch (trained by Matt Buchanan) and Dr. Church; Judith and Brown Prince, both saddle class horses preparing for the Wissahickon Horse Show; Westchester and Geiger, both jumpers; and King Carter, Woodrule, Glen L., and Major Wind, all steeplechasers.[67]

Horse training and breeding, however, were not the only activities at Erdenheim under Carson's ownership. Beginning in 1904 Erdenheim's gardener, Frank Smith—under the direction of Carson himself—began raising freshwater terrapin at the farm. Beginning with a concrete-lined cistern under the protective shell of the greenhouse, Smith quickly raised more than one hundred terrapin. These were soon transferred to a terrapin pond, probably the one built by Carson just to the southeast of "New" Erdenheim. Carson's motivation for this well-regarded achievement was made clear by the press in 1909, however: "Mr. Carson was delighted with the ease with which the delicacy was provided, fresh and finely flavored, for his household."[68] Prize sheep were also a perennial production of Erdenheim's fields.[69]

FACING: "On the Wissahickon. Green Houses Erected by Robert Carson at Erdenheim. Cost: $20,000" (DAMS 469), circa 1903, Thomas H. Shoemaker Germantown and Philadelphia Portraits and Views Collection (V86), folio 4, Historical Society of Pennsylvania.

Fires and Thefts at Erdenheim

The threat of conflagration in the barns and stables at Erdenheim had always been a concern, but under Robert Carson's ownership, such events proved devastating. The first of these fires under Carson's watch occurred in 1902, when flames gutted two stables, the 1870 Welch barn, and an indoor quarter-mile track, inflicted $100,000 in damages, and killed five thoroughbreds, including Theodore Cuyler Patterson's polo horse Kasim (incorrectly listed as "Hasin" in some accounts), "which ranked first in the official pony stud book of England and was considered the best pony of its class in America." Smoke was discovered pouring from the two-story stallion house by a night watchman around 2:00 A.M., then quickly spread to nearby buildings despite the efforts of a bucket brigade. R. Nelson Buckley, Carson's nephew and an accomplished polo player, came over from his nearby summer residence ("Old" Erdenheim) with his houseguest, George Newhall, to help rescue horses and extinguish the flames. Together with John McCloskey and his son James, the men succeeded in getting all the horses out of the stables. As the freed horses galloped up and down Stenton Avenue (then Blue Bell Road), someone from the crowd gathered to watch the spectacle wandered into the enclosure, opened the gate, then opened the doors to the Welch barn just as Kasim, two colts owned by McCloskey, and two others owned by Cyrus Lukens and John Deignan, entered the enclosure. In the midst of the conflagration surrounding them, the horses took the open doors as a cue and raced back into the burning barn, where all five burned to death.

Newhall, both McCloskeys, and a groom, Daniel Morning, suffered serious burns for their second efforts to stop the horses from reentering the barn. The stallion house, the covered ring stable with exercise track, and the 1870 Welch barn (with the exception of its stone walls) were all destroyed in the fire.[70]

As under Kittson's ownership, fires seemed to recur every few years. A fire during the night of September 17, 1909, for example, inflicted heavy damage at the farm totaling $60,000, entirely destroying the main stables and killing four horses, three polo ponies, and a prize-winning heifer.[71] This stable was likely the "huge stone barn, with its red roof," located north of "Old" Erdenheim in 1902.[72] But fires were not the only concern at Erdenheim. Intrepid thieves made off with "Knightbird," a highly prized hunting horse valued at $1,000 and belonging to Clarence Kline, from its stable at Erdenheim on August 6, 1913. Two months earlier thieves had also made off with two workhorses at the farm.[73]

Death of Robert Carson

Robert Carson's demise on October 15, 1907, came under rather exceptional and shocking circumstances. As the first act of *The Merry Widow* at the Chestnut Street Opera House came to its conclusion, patrons heard a cry erupt from the vicinity of Carson's seat that seemed to mimic the last few words of the act. Mrs. Carson and Carson's brother-in-law, John D. Parsons, tried to revive him without success. To alleviate embarrassment Mrs. Carson and several ushers pretended that Carson had simply fainted. Once the second act began and the theater was again dark, two ushers came and removed his body; an ambulance then transported Carson's body to the apartment house he owned on Walnut Street.[74] In the days following his death, information emerged confirming that he died from "organic heart disease," from which he had been suffering since 1903, and that his personal physician had explicitly advised him not to attend the show that night, on account of "an acute attack of indigestion" he had suffered that week.[75]

Within only a couple of weeks of Carson's death, news emerged about the terms of Carson's will. Naming the Real Estate Title and Trust Company the chief executor of his estate, Carson's will called for $1,000,000 to be spent on the erection of buildings at a new orphanage for girls that would be called Carson College, with much of the balance of his estate to be left to this school as an endowment. While Carson had developed the idea of such a school in large part as an inspired reaction to Stephen Girard's establishment of Girard College—an orphanage for boys—at Philadelphia in 1848, he also intended Carson College to be something of a monument to Robert Carson the man. "He talked to me about a girl's college," his widow said in 1911, "and said that he wanted it erected as a monument to himself, because, as he said, he had sweated blood in getting what he had."[76]

As one might expect, the Carson heirs were not exactly thrilled to learn that Carson had bequeathed a substantial part of Erdenheim and $4,000,000 to establish a girl's orphanage. To block the bequest, Carson's heirs filed appeals against the probation of the will, citing their belief that such a school was not only impractical but also unnecessary. After the local judge dismissed their objections, the heirs appealed unsuccessfully to the state supreme court another four times.[77] The result was a protracted legal battle that produced an exhaustive accounting of the estate while it was tied up in legal battles. While it lies beyond the scope of this history of Erdenheim, these accounting papers accompanying his will record provide a fascinating glimpse of Carson's investments, his expenses, his employees, and even the obligations to the stock farm and the use of the land at Erdenheim.[78] Initial plans, curiously, were to use the entirety of Erdenheim's then-350 acres for the proposed orphanage, but Carson's trustees ultimately determined that only 100 acres would be needed for the school.[79]

Whitemarsh Valley Hunt Club

A few months before Carson's death in 1907, Carson also extended a long-term, ten-year lease to the Whitemarsh Valley Hunt Club on a property located along West Mill Road, just to the north of the one-mile oval.[80] Organized in 1904, with Robert Carson as one of its earliest members, the Whitemarsh Valley Hunt Club began offering races at Erdenheim almost immediately.[81] By 1910, however, members of the club, which had begun using other facilities at Erdenheim for its meetings and its hunts, began to grow anxious about what would become of Erdenheim in the wake of Carson's death. R. Nelson Buckley, Carson's nephew and "one of the most enthusiastic horsemen of the club," joined other members in advocating for the establishment of a new stock farm in the vicinity. To motivate interest from non–club members who owned farms nearby, the members of the club held their annual community breakfast at Erdenheim, foregoing their own familiar elite status and extending it to their guests by waiting tables and hosting various sporting affairs, including "half-mile races for purses for farmers' horses, potato and sack races for the sons of farmers, and finally a free-for-all jumping contest."[82]

During the typical hunt season, meanwhile, the White Marsh Valley Hunt Club held "drag runs" every Wednesday morning, Saturday afternoon, and holiday. Such events usually drew between twenty-five and fifty participants, and

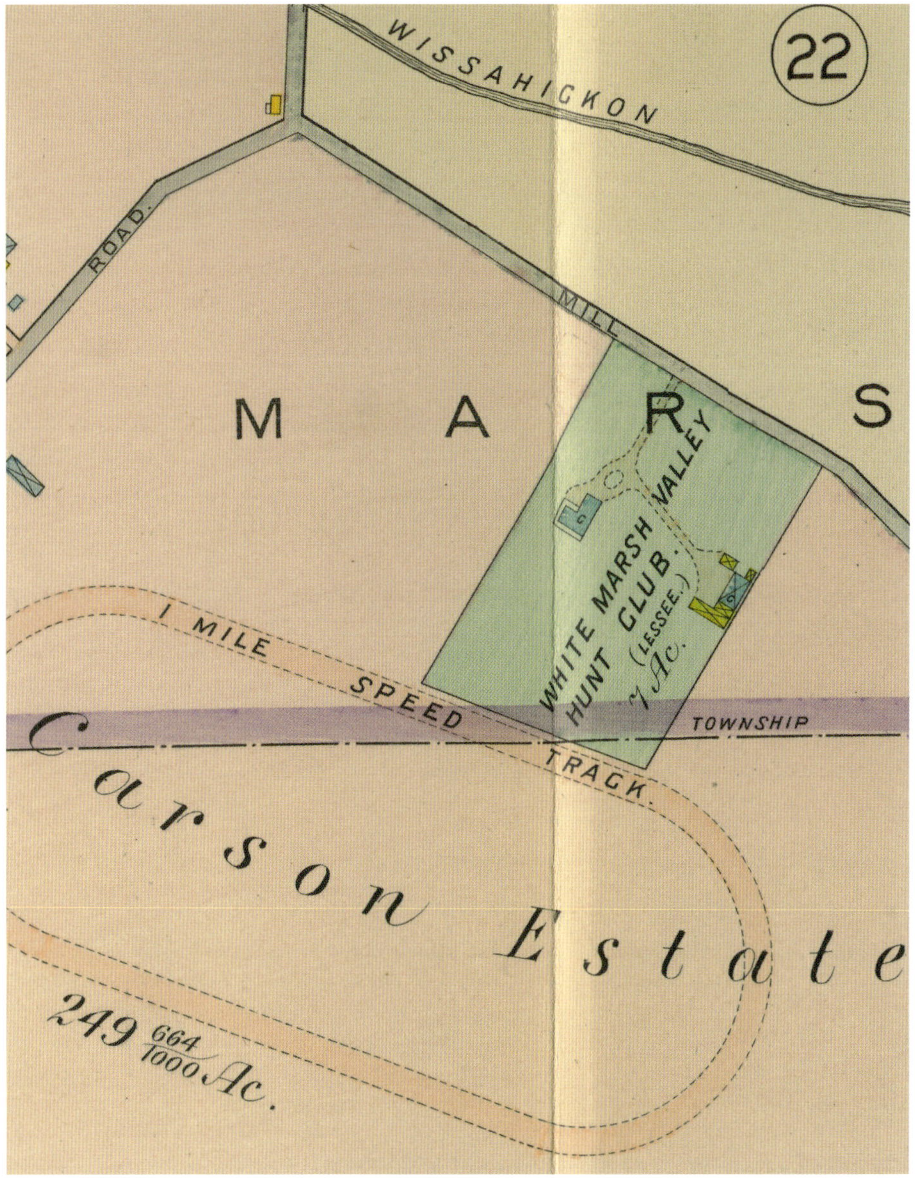

White Marsh Valley Hunt Club parcel as of 1909, from *Atlas of Properties on the Reading Railway,* 1909, plate 20, Map Collection, Free Library of Philadelphia. Carson issued a ten-year lease on the parcel to the club in 1907.

the leased grounds of the hunt club included a stable and kennel complex at the east end of the property, plus a rather large club house on the ridge just above Erdenheim's one-mile track.[83] Whether or not this clubhouse is the same one referenced in descriptions of the races at Erdenheim during this period is not known with certainty, but it seems likely given that no other clubhouse is known to have been built during this period.

Carson College

For five years following Carson's death, activity at Erdenheim remained in limbo, its agents and boarders still making the most of the stables and grounds there, while Carson's heirs and many of Erdenheim's most frequent visitors tried to make sense of what would happen to the beloved farm once Carson's estate was settled. The death of Carson's widow in 1912 freed up the estate for disposition, leaving roughly $500,000 to be distributed among Mrs. Carson's next of kin (the Carsons had no children), another $1,000,000 originally designated by Carson as a life estate, now to be redirected for the construction of Carson College, and another $5,000,000 as an endowment to that institution.[84] Construction of the new Carson College drew a great amount of positive press, with many heralding the imaginative design and layout for the school. Architect Albert Kelsey's plans called for more than twenty buildings and a massive "windbreak" of trees surrounding the entire property. On the northwest side (adjoining Erdenheim), Kelsey planned various sylvan features, including "The Moon Well" and "The Star Well," joined to one another by a long tunnel of trees known as "The Wonder Walk." Other features were designed to encourage Carson College's children to play among and simultaneously learn about the trees and forest creatures, including "The Peter Pan Puzzle" and "The Bird Walk."

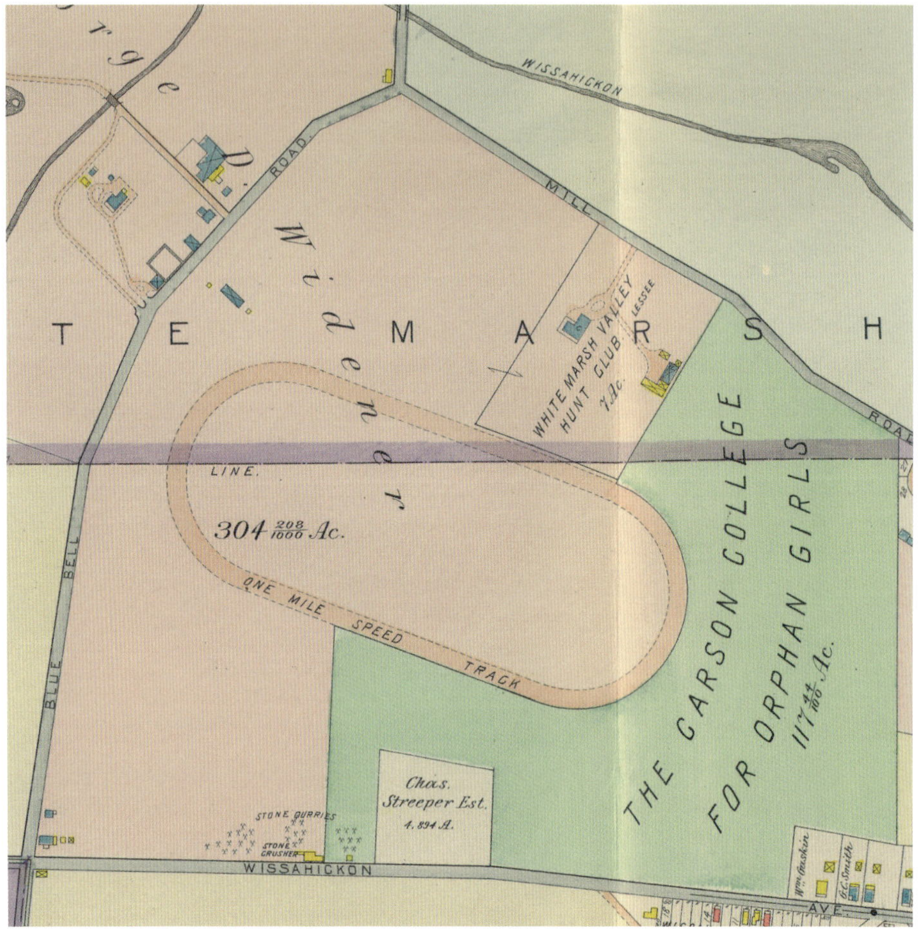

Division of the Carson lands showing the new land set aside for Carson College, 1916. Detail from *Atlas of the North Penn Section of Montgomery County*, 1916, plate 20, Map Collection, Free Library of Philadelphia.

Each of the cottages, meanwhile, was intended to serve as both home and school to the girls and was equipped with a small kitchen in addition to the classrooms, dorm rooms, and living spaces. Kelsey intended these buildings to serve as an "anthology," each possessing the name of a particular flower and representing that flower's "concept." Together these designs represented "an allegorical portrayal of a woman's life from her own birth to the time that she gives birth to another little life." The various stages of his allegory bore their own titles, including "Childhood and Play," "Girlhood and Study," "Maidenhood and the Melody of Life," and "The Glory of Motherhood and the Mystery of Time and Space." A memorial tower to Carson at the center of campus employed its own allegorical theme, one that required climbing the tower to fully appreciate the design.[85]

To some outsiders the girls at the Carson School led enviable lives. Unlike their peers elsewhere, the girls at Carson were encouraged to dress individually rather than in uniform, frequently received ornate holiday gifts, and were permitted to celebrate Halloween festivities in the tunnels connecting the various cottages on campus. In addition to their conventional classroom studies, the Carson girls also learned weaving, spinning, and gardening, and they helped maintain a working dairy farm on site.[86] Carson College remains in operation today, albeit with a slightly modified mission, under the name of Carson Valley Children's Aid.

The success of Carson College's completion aside, the fate of the remainder of Erdenheim remained deeply uncertain in 1916. Many feared that the fabled horse farm had finally met its end, and that the best efforts of Erdenheim's most loyal patrons would not be sufficient to save the land from development or distribution into dozens of smaller parcels. In some respects, the fate that emerged was a mixed blessing, one that kept the name of Erdenheim alive and in fact enhanced its reputation, while also resulting in the destruction of many of Erdenheim's most famous and beloved landmarks. The story of Erdenheim's twentieth century, then, is not unlike that of the caterpillar and the butterfly—a casting off of the old, a lengthy and painful transformation, and the emergence of something far greater and more beautiful than its progenitor.

Notes

1. Untitled blurb, *Turf, Field, and Farm,* May 26, 1882, 351, and "The Sale of Erdenheim." The latter quoted material is from an anonymous Turf Club member quoted in the article. See also "The Purchase of Erdenheim by Commodore Kittson," *St. Paul Daily Globe,* May 29, 1882, 1.
2. Untitled column, *Spirit of the Times,* May 19, 1888, 620.
3. "Com. Kittson's Latest," *St. Paul Daily Globe,* January 13, 1882, 2. The article incorrectly reported the sale price of Erdenheim at $150,000.
4. "The Purchase of Erdenheim by Commodore Kittson," *St. Paul Daily Globe,* May 29, 1882, 1; Henry C. Klassen, "Kittson, Norman Wolfred," *Dictionary of Canadian Biography,* volume 9, 2000, http://www.biographi.ca/009004-119.01-e.php?&id_nbr=5627; and C. W. Rife, "Kittson, Norman Wolfred," *Dictionary of American Biography,* volume 10, Dumas Malone, ed. (New York: Charles Scribner's Sons, 1933), 441. For a lengthy treatment of Kittson's early career in Minnesota Territory, see Clarence W. Rife, "Norman W. Kittson, a Fur-Trader at Pembina," *Minnesota History* 6 (September 1925): 225–52.
5. Wiley, "Louis Kittson," *Biographical,* 364–65. Louis Kittson should not be confused with another Kittson son, Louis Coyle Kittson, who died in 1892 at the age of twenty-five. See "The Death Roll," *Duluth News-Tribune,* December 4, 1892. Commodore Kittson named eleven living children in total in his will, as well as one deceased son.
6. "Millionaire Horsemen," *St. Paul Daily Globe,* April 23, 1884, 3; and Rife, "Kittson."
7. "Com. Kittson's Latest," *St. Paul Daily Globe,* January 13, 1882, 2. In a snarky and remarkably unkind swipe at Welch, the writer of this article enumerated the various brood mares included in the sale, concluding the lengthy list with "Henrietta Welch, the dam of gossip." While Henrietta Welch was a mare owned by Welch, and her dam was a horse named Gossip, the failure to capitalize the latter term seems too coy to have been accidental, especially considering Welch's already tense relationship with the press.
8. "The Erdenheim Sale," *New York Herald,* November 5, 1882.
9. "The Erdenheim Sale," *New York Herald,* October 17, 1882.
10. Lewis A. Lukens, et ux., to Norman W. Kittson, May 21, 1883, Deed Book 273, page 333, Historical Society of Montgomery County.
11. "Erdenheim on Wissahickon," *St. Paul Daily Globe,* December 18, 1882, 2; Untitled column, *Spirit of the Times,* May 19, 1888, 620; and Untitled column, *Spirit of the Times,* September 7, 1889, 274. Daniel W. Woodmansee, a long-time associate of Kittson's, ran his operations at Midway. See Untitled column, *Spirit of the Times,* May 19, 1888, 620.

12. "Erdenheim on Wissahickon," *St. Paul Daily Globe,* December 18, 1882, 2.
13. "Erdenheim on Wissahickon," *St. Paul Daily Globe,* December 18, 1882, 2; and "Flames at Erdenheim," *New York Times,* October 1, 1884.
14. "Erdenheim on Wissahickon," *St. Paul Daily Globe,* December 18, 1882, 2.
15. Bean, *History,* 1146; and *Atlas of the Properties on the Reading Railway, Embracing Cheltenham, Abington, Springfield, and Parts of Moreland and Whitemarsh Townships,* surveys by Ellis Kiser and J. M. Lathrop (Philadelphia: A. H. Mueller, 1909), Map Collection, Free Library of Philadelphia; "Flora Temple's Grave," *Philadelphia Record,* unknown date (c. 1888), clipping in the private archival collection of Mary Jane Echenhofer.
16. "The Erdenheim Stables," *St. Paul Daily Globe,* December 15, 1884, 2.
17. "Mr. Kittson to Withdraw," *St. Paul Daily Globe,* August 30, 1886, 9.
18. Untitled column, *Spirit of the Times,* November 3, 1888, 560.
19. Untitled column, *Spirit of the Times,* November 3, 1888, 560.
20. Untitled column, *Spirit of the Times,* November 3, 1888, 560.
21. "Millionaire Horsemen," *St. Paul Daily Globe,* April 23, 1884, 3. This article includes a brief interview with Splan, in which he explains how he got his start as a driver.
22. "Millionaire Horsemen," *St. Paul Daily Globe,* April 23, 1884, 3.
23. "The Erdenheim Stables," *St. Paul Daily Globe,* December 15, 1884, 2.
24. "Erdenheim: Fire at the Famous Stock Farm of Commodore Kittson, near Chestnut Hill," *Philadelphia Inquirer,* October 1, 1884, and "Flames at Erdenheim," *New York Times,* October 1, 1884.
25. "Erdenheim: Fire" and "Flames at Erdenheim." The *Inquirer* reported the loss as $13,000, which seems low.
26. "The Erdenheim Stables," *St. Paul Daily Globe,* December 15, 1884, 2.
27. "Mr. Kittson's Stable Burned," *New York Times,* October 16, 1885.
28. Billy Young, "Speaking of Men and Horses," *Cleveland Gazette,* November 18, 1944. For more on Healey's early career, see The Judge, "Tom Healey, Trainer," *Turf,* November 15, 1913.
29. Untitled blurb, *New Hampshire Patriot,* March 4, 1886.
30. "Mr. Kittson to Withdraw," *St. Paul Daily Globe,* August 30, 1886, 9.
31. "The Erdenheim Sale: Commodore Kittson's Horses at Auction and Fair Prices Obtained," *New York Herald,* October 7, 1886; Untitled column, *Spirit of the Times,* September 7, 1889, 274.
32. "Death of Norman W. Kittson," *New York Times,* May 11, 1888; "Death of Norman W. Kittson," *Turf, Field, and Farm,* May 18, 1888, 421; and Untitled column, *Spirit of the Times,* May 19, 1888, 620.
33. "A Pioneer of Minnesota," *New York Times,* May 20, 1888; and Untitled column, *Spirit of the Times,* May 19, 1888, 620.
34. "The Rome of Racers: Erdenheim's Famous Horses Sold at Public Auction," *Philadelphia Inquirer,* November 9, 1888; "Sale of the Erdenheim Stud," *Daily Inter Ocean,* November 9, 1888; "Sale of the Erdenheim Stud, Property of the Kittson Estate," *Kentucky Livestock Record,* November 17, 1888, 306; and "The Erdenheim Sale," *Turf, Field, and Farm,* November 16, 1888, 460.
35. "Yearlings of High Degree," *New York Herald,* May 14, 1889.
36. "The Erdenheim Yearlings," *Turf, Field, and Farm,* April 11, 1890, 417.
37. "A Big Case Settled," *Philadelphia Inquirer,* September 26, 1889, 3. For the full text of the Supreme Court appeal, see *Farley v. Kittson,* 120 U.S. 303 (1887).
38. "Last Will and Testament of Norman W. Kittson," August 26, 1889, Will Book 19, page 504, Montgomery County Archives. For the Carson deed transaction, see James E. Kittson, et al., to Robert N. Carson, July 6, 1893, Deed Book 385, page 253, Montgomery County Archives. Inexplicably the calculation of acreage in this deed equals 148.648 acres, nearly an acre more than the Lukens to Kittson transaction, even though the parcels are described as being identical. This difference may be the result of additional shifts in the riverbed of the Wissahickon, which was often used to define the boundary between these parcels.
39. "Erdenheim Stock Farm," *Sacramento Record-Union,* June 8, 1893, 1; and "An Old Stock Farm to Be Revived," *Philadelphia Inquirer,* June 7, 1893. The first article incorrectly reports one of the tracts as being 111.67 acres rather than 57.
40. James E. Kittson to Louis Kittson, November 24, 1893, Deed Book 389, page 365, Montgomery County Archives.
41. "Kittson Was Not Married," *New York Times,* March 4, 1896.
42. "Norman A. Kittson's Will," *New York Times,* May 20, 1888. The title is clearly in error but refers to Erdenheim's Kittson.
43. "Notes about Horses and Trainers," *New York Herald,* December 17, 1891, 3; "Erdenheim Stock Farm," *Philadelphia Record,* unknown date (c. 1903), clipping in the private archival collection of Mary Jane Echenhofer.
44. Agreement, Louis Kittson, et ux., to Robert N. Carson, April 27, 1896, Deed Book 412, page 486, and Louis Kittson, et ux., to Robert N. Carson, June 12, 1896, Deed Book 415, page 1, Montgomery County Archives. See also "Erdenheim Sold: The Home of Famous Leamington Bought By Robert N. Carson," *Philadelphia Inquirer,* April 30, 1896. It should be noted here that two maps from the period errantly suggest that the original "Old" Erdenheim parcel was for a time owned by H. L. Lipman—almost certainly Hymen Lipman, a Philadelphia entrepreneur who had made a fortune in the postcard business during the 1860s and 1870s, and multiplied that fortune with his patent of the first pencil with an attached eraser. But Lipman was dead by November 1893, and his only son was Lewis H. Lipman. Other cadastral maps of the same period show "Mrs. Lehman"—most likely a reference to Lipman's widow, Mary Lehman Lipman—as the owner of the parcel immediately southeast of "Old Erdenheim." Carson is shown as owner of "Old" Erdenheim in plate 18 of Ellis Kiser, Otto Barthel, and B. A. Skeels, *Atlas of Cheltenham, Abington, and Springfield Townships and Vicinity, Montgomery County, Pa.* (Philadelphia: A. H. Mueller, 1897). This same map shows "Mrs. Lehman" as the owner of the 30-acre parcel immediately southeast of "Old" Erdenheim. For an overview of Lipman's life, see "Obituary: Hymen L. Lipman," *The American Stationer,* November 16, 1893, 992–93.
45. "Suicide of James E. Kitson [*sic*]," *Washington Times,* November 17, 1900, 8.
46. "Story of the Globe," *St. Paul Daily Globe,* December 21, 1902, 23.
47. "Civil Service Exam Today," *Philadelphia Inquirer,* May 19, 1920, and "Difenderfer May Be Bowled Out," *Philadelphia Inquirer,* December 28, 1913.

48. "Dies in Theatre, Audience Thinks It Fainting Fit," *Philadelphia Inquirer,* October 16, 1907; "Drops Dead in Theatre," *New York Times,* October 16, 1907. Carson's middle name is rendered as Neidermark or Niedermark in some sources. Other sources place his date of marriage at January 27, 1874, but without documentation, instead of 1894.
49. David Contosta, *Philadelphia's Progressive Orphanage* (University Park: University of Pennsylvania Press, 1997), 8–9. Contosta's book is an outstanding study of Carson College, its origins and evolution, and its overall history. Even Contosta, however, was stunned at the dearth of information on Carson's private life.
50. Whitemarsh, Montgomery County, Pennsylvania, *1900 United States Federal Census,* roll 1445, 13A, National Archives and Records Administration, Ancestry.com.
51. "Newport," *Philadelphia Inquirer,* August 25, 1901, and "Accident on the Marietta," *Sun,* August 29, 1905, 10.
52. "Erdenheim," *St. Paul Globe,* August 24, 1896.
53. "Where Robert N. Carson's Millions Will Rear a College for Orphan Girls," *Public Ledger,* October 24, 1907. See also Whitemarsh, Montgomery County, Pennsylvania, *1900 United States Federal Census,* roll 1445, 13A, National Archives and Records Administration, Ancestry.com.
54. "Erdenheim," *St. Paul Globe,* August 24, 1896.
55. "Erdenheim," *St. Paul Globe,* August 24, 1896.
56. "All the Turf News: Stryke Takes the Erdenheim Welter Handicap at Morris Park," *Philadelphia Inquirer,* June 8, 1890.
57. C. Edward Sparrow, "Whimsy Gets Erdenheim," *Sun,* October 1, 1915.
58. "Girard and His Gift Swayed Carson," *Philadelphia Record,* October 27, 1907.
59. "Erdenheim Events Tomorrow," *Philadelphia Inquirer,* September 7, 1906; "Society Out for Erdenheim Races," *Philadelphia Inquirer,* September 9, 1906, 12.
60. "Erdenheim Track Races," *Philadelphia Inquirer,* October 20, 1904. For the results of another such race, see "Pete Wins Match Race with Thos. K," *Philadelphia Inquirer,* November 8, 1908.
61. "Racing at Erdenheim," *Philadelphia Inquirer,* August 29, 1909.
62. "Girard and His Gift Swayed Carson," *Philadelphia Record,* October 27, 1907.
63. These changes are visible on Ellis Kiser and J. M. Lathrop, surveyors, *Atlas of Properties on the Reading Railway, Embracing Cheltenham, Abington, Springfield, and Parts of Moreland and Whitemarsh Townships* (Philadelphia: A. H. Mueller, 1909), plate 20. Compare this map, for example, with *Atlas of the Properties in the North-West Suburbs of Philadelphia* (Philadelphia: G. Wm. Baist, 1893), plate 10, which shows only the main house and two smaller structures nearby.
64. "Erdenheim Farm Familiar Landmark," *Philadelphia Inquirer,* May 9, 1906; Advertisement, *Philadelphia Inquirer,* May 13, 1906, 13.
65. Theodore Cuyler Patterson, "Breeding Thoroughbred Ponies," *Outing, an Illustrated Monthly Magazine of Recreation,* December 1900, 274.
66. Williams, the McCloskey estate, and one of their boarders were the litigants in a protracted case that reached the Pennsylvania Superior Court, involving the recovery of the cost of boarding for a horse kept at Erdenheim by the defendant, who had actually instructed that the horse be killed. See *Williams v. Patterson,* 38 Pa. Superior Ct. 322 (1909), *Pennsylvania Superior Court Reports,* volume 38 (1909): 322–26.
67. "Erdenheim Farm Familiar Landmark," *Philadelphia Inquirer,* May 9, 1906, 10.
68. "Ra[i]sing Terrapin," *Patriot,* August 30, 1909, 5.
69. "Girard and His Gift Swayed Carson," *Philadelphia Record,* October 27, 1907.
70. "Erdenheim in Flames," possibly *Philadelphia Record,* unknown date but probably June 8, 1902, clipping in personal archival collection of Mary Jane Echenhofer; "Thoroughbreds Go Up in Smoke," *Philadelphia Inquirer,* June 8, 1902. Carson had leased the farm itself to John McCloskey. This article incorrectly reports McCloskey's name as "John McCluckey." For more on Kasim, see Theodore Cuyler Patterson, "Breeding Thoroughbred Ponies," *Outing, an Illustrated Monthly Magazine of Recreation,* December 1900, 270–77.
71. "The Man on the Corner," column clipping from unknown newspaper, September 23, 1909, Springfield Township Historical Society, Folder 2008.96.2.
72. E.M., "Local History Sketch," *Ambler Gazette,* October 16, 1902, 2.
73. "Valuable Hunting Horse Stolen from Stock Farm," *Philadelphia Inquirer,* August 7, 1913.
74. "Dies in Theatre, Audience Thinks It Fainting Fit," *Philadelphia Inquirer,* October 16, 1907.
75. "Carson Disobeyed Physician's Orders," *Philadelphia Inquirer,* October 17, 1907.
76. "Robert N. Carson," *Patriot,* March 11, 1911, and "Girard and His Gift Swayed Carson," *Philadelphia Record,* October 27, 1907.
77. Toll and Schwager, volume 2, 959. For an example of newspaper coverage of the battle over the will, see "Fight for Income of $4,000,000 Estate," *Philadelphia Inquirer,* March 26, 1909. For local court coverage over the trial regarding settlement of the will, see *Montgomery County Law Reporter,* volume 28, 1912, 39–52.
78. Estate of Robert N. Carson, Register of Wills File 25769, Montgomery County Archives.
79. "Stock Farm Sold at Chestnut Hill," *Philadelphia Inquirer,* April 22, 1916.
80. "Where Robert N. Carson's Millions Will Rear a College for Orphan Girls," *Public Ledger,* October 24, 1907.
81. "Town and Country Calendar," *Town and Country,* September 10, 1904, 3.
82. "Hunt Club at Erdenheim Serves Spring Breakfast to Montgomery County Farmers," *Philadelphia Inquirer,* March 27, 1910. Buckley's name is reported in several variations in the article, but none of them as Robert Nelson Buckley, his actual name. Buckley, an avid sport hunter, must have had far too much leisure time on his hands. The article reported that he had already hunted ninety-one times that year, "a record for an American, and he is striving to round out an even hundred times of following the hounds before the season is finally over."
83. "Masters of the Hunt," *Town and Country,* November 9, 1907, 20.
84. "Mrs. R. N. Carson Buried," Philadelphia Inquirer, July 7, 1912.
85. "Carson College Will Be Real Fairyland," *Ambler Gazette* [?], May 1917 [?], and M'Liss, "Architect Realizes Dream in Corson [*sic*] College for Girls," *Ambler Gazette* [?], May 1917 [?], Clippings file, Historical Society of Montgomery County.
86. Toll and Schwager, *Montgomery County,* vol. 1, 625–26.

CHAPTER FIVE

George D. Widener Jr., A Man of Titanic Influence

When George D. Widener Jr. purchased Erdenheim in 1916, the famous old stud farm had been in a steady period of decline. Under both Commodore Kittson and Robert N. Carson, the breeding operations were no longer turning out champions at anywhere near the success rate that Aristides Welch had for twenty years, and many of its boarders were owned by lessees rather than the farm itself. John McCloskey, Welch's masterful groomsman and trainer who had stayed on to assist the Kittson and Carson families, was dead. America's best horsemen still stabled their horses at Erdenheim, and the Whitemarsh Valley Hunt Club routinely traipsed across its fields on sporting days, but the upcoming expiration of the hunt club's lease and the terms of Carson's will threatened to derail all of that, too. Erdenheim, it seemed, needed a savior, and everyone close to Erdenheim seemed to know it.

George D. Widener Jr.

The Widener family's ancestors first came to Philadelphia aboard the *Phoenix* from Rotterdam (via Portsmouth, England) on November 2, 1752, and settled in Northern Liberties by 1774. The great-grandfather of George D. Widener Jr., John (Johannes) Widener, was born at Philadelphia in 1790, and by 1813 he was heavily involved in the brick manufacturing business in Philadelphia. One of his two sons, Peter Arrell Brown (P. A. B.) Widener was born at Philadelphia in 1834, and it was under his banner that the Widener family made the first of its fortune. Widener's grandfather had made his money the old-fashioned way—through hard work, risk taking, and rampant speculation. Beginning with a butcher shop in the Spring Garden Market and other shops scattered around Philadelphia, the Widener clan seemed, like their contemporary Aristides Welch, to always be at the right place at the right time.

The American Civil War brought opportunity for many a wise businessman, and the Wideners did not miss that boat either, profiting heavily on the sale of mutton to the Union army. No doubt their reputations as influential Republicans and supporters of Lincoln opened doors not otherwise accessible to others. At first a major figure in the meat-packing business in the city, P. A. B. Widener later turned his sights to the street railways and other major industrial concerns, so that by the end of his career, he had served as a director of the Metropolitan

FACING: Erdenheim cattle graze on the hillside below the graves, now unmarked, of Whitemarsh's earliest settlers. Photo by Eric Plaag, May 2012.

MR. LASKER

Street Railway Company of New York, the Philadelphia and Reading Coal and Iron Company, the Land Title and Trust Company, United States Steel, and International Mercantile Marine. With the fortune P. A. B. Widener made, he built two schools for handicapped children in Philadelphia and New Jersey, and he donated the building that today houses the Central Branch of the Free Library of Philadelphia. He also served a brief stint as the treasurer of Philadelphia.[1]

In June 1861 P. A. B.'s son George Dunton Widener was born, and like his father, he got his start in food, working for many years at a grocery store before entering his father's streetcar business. Within just a few years, he was recognized as an expert on traction transportation, and there is little doubt that he must have known Robert N. Carson, his contemporary and competitor, during the late nineteenth century. In addition to his co-ownership of the Philadelphia Traction Company and nearly a dozen smaller, often local railway lines, George Dunton Widener invested heavily in American Tobacco Company and U.S. Steel, and he served as a director in numerous other companies, many of them closely connected with the needs and demands of the traction industry. In November 1883 he married Eleanore Elkins, the daughter of William Lukens Elkins, another wealthy Philadelphia scion who got his start in the grocery business, made a fortune pioneering the process of oil refinement and subsequently partnering with Standard Oil, then partnered with P. A. B. Widener on numerous projects in the traction business. George and Eleanore had three children before 1892—Harry Elkins Widener, George Dunton Widener Jr., and Eleanor Widener.[2]

FACING: Jessie Sloane Widener and George D. Widener Jr., undated but circa 1940, Cook Collection 2428, Keeneland Library.

George Dunton Widener Jr. was born on March 11, 1889, and after securing his education at the Delancey School in Philadelphia, his path was a different one than that of his forebears. Until he was well into his twenties, the young Widener lived at Lynnewood Hall, the massive Elkins Park mansion designed by Horace Trumbauer that had housed the widower P. A. B. Widener and the families of both of his sons—Joseph E. and the elder George—since 1900.[3] He did not attend college, and rather than having to make his own fortune, he was assigned the task of managing the one his family already possessed. He also served as a director of the Land Title Bank and Trust Company and the Electric Storage Battery Company of Philadelphia.[4] For all the family wealth, though, George D. Widener Jr. would have to suffer through tragedy as a young man. In April 1912, as news shocked the world about the sinking of the *Titanic* in the North Atlantic, young Widener learned that both his father and his brother, Harry Elkins Widener (after whom Harvard's main library was later named), had gone down with the ship. His mother, Eleanore Elkins Widener, was fortunate to find her way into a lifeboat.[5] Two months later, in spite of this tremendous loss to the family, George's sister Eleanor quietly wedded Fitz Eugene Dixon (1888–1982).[6]

The death of Widener's father and brother created something of a sea change in young George's life. The following year—prior to his purchase of Erdenheim—Widener entered the world of horseracing for the first time, purchasing the stallion Garry Herrmann and several mares. To facilitate his arrival in this field, he immediately registered his racing silks, whose light blue and dark blue hoops and dark blue cap would become familiar and iconic to the racing crowd over the next sixty years. Two years later his two-year-old filly Columbine was regarded as the best of the year. His uncle Joseph E. Widener, long a powerful figure in the racing world, no doubt aided George Widener's arrival on the racing scene.[7] With these early successes in racing and an unimaginable fortune available to him, it was only a matter of time before Widener found a place close to his family's home with which to build a racing empire. Asked many years later to explain his status as a god among horsemen, Widener humbly acknowledged that this path was not necessarily an obvious one for him: "I did not seek it, but once that was the fact, it seemed to be a useful and creditable way of life."[8]

George D. Widener Jr. Purchases Erdenheim

Acquiring the historic Erdenheim parcels must have been fairly easy for Widener to pull off. As a horseman well acquainted with the members of Whitemarsh Valley Hunt Club and a resident of nearby Elkins Park (just seven miles to the southeast of Erdenheim), it seems likely that Widener was biding his time until the Carson heirs—and most particularly Carson's nephew R. Nelson Buckley, who was a key figure at the Whitemarsh Valley Hunt Club—were ready to sell the remaining land not deeded over to Carson College. On May 15, 1916, Widener purchased much of the old Carson land, specifically tracts D through L of the estate as it had been divided for settlement purposes, totaling 155.56 acres and representing the remainder of "Old" Erdenheim. For this plot Widener paid just $1.00, raising obvious questions about whether Widener and Buckley had some other side agreement regarding compensation for this valuable land.[9] That same day Widener also purchased from the Real Estate Title Insurance and Trust Company another 148.648 acres of land on the west side of the Wissahickon, representing the old Lukens parcel (or "New" Erdenheim) that had been owned by Kittson and Carson, for the sum of $61,200. This purchase excluded a parcel

The Widener family crest, from Thomas H. Bateman, compiler, *Widener-Dunton-Elkins-Broomall and Allied Families,* 1953.

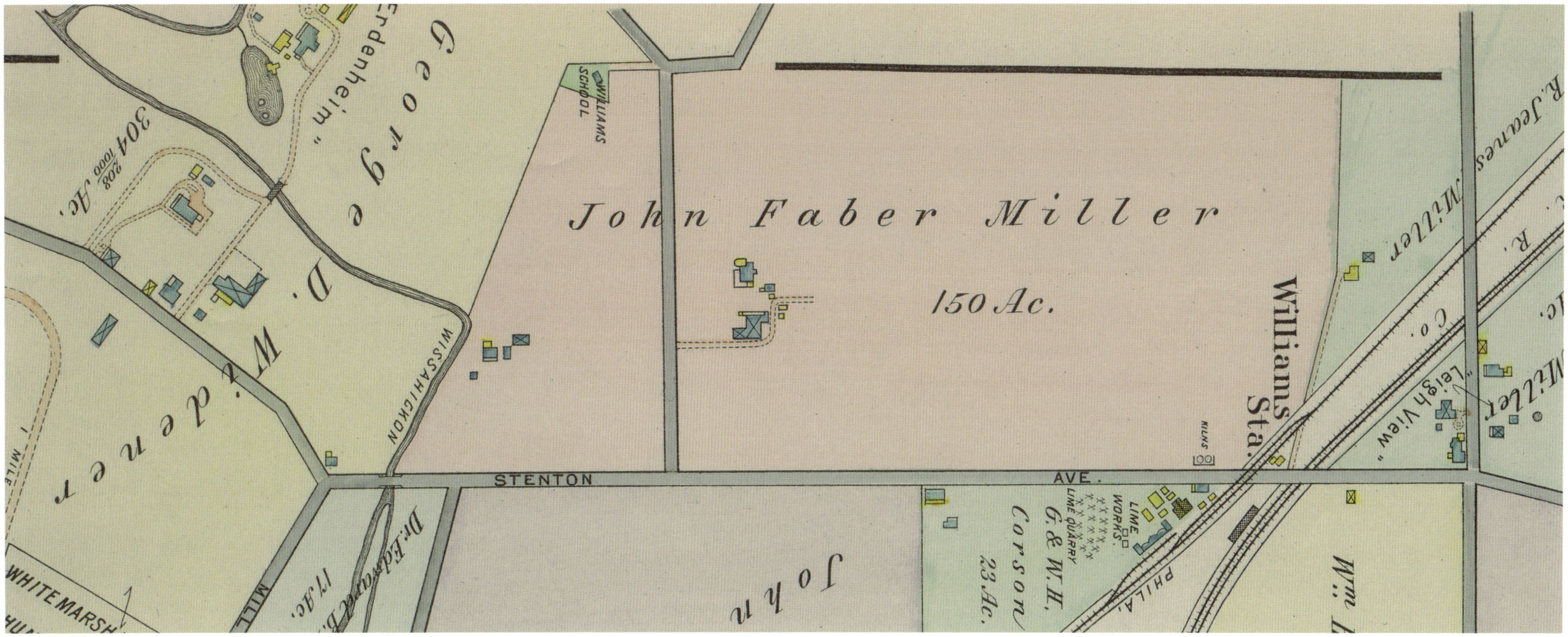

The old Daniel Williams parcel (labeled "John Faber Miller") as it appeared in 1916, immediately prior to its transfer to Widener. Detail from A. H. Mueller, *Atlas of the North Penn Section of Montgomery County*, 1916, plate 22.

of just over 16 acres previously subdivided to the Carson estate and ultimately transferred to the Whitemarsh Valley Country Club.[10]

From the beginning Widener intended to restore Erdenheim's nineteenth-century reputation as a first-rate stock farm. Accomplishing that task, however, meant consolidating the land surrounding the old Carson estate. Among his first targets was the substantial 150-acre Williams parcel on both sides of lower Flourtown Road and west of Stenton Avenue, still owned by the heirs of Daniel

Williams. To accomplish this, Widener appears to have secured the services of a neighbor, John Faber Miller, to act as a surrogate. For the price of $1.00, Miller purchased 149 acres, sixty perches of the old Williams parcel on September 13, 1916. Just five days later, Miller flipped the entire parcel to Widener for the same price of $1.00. These transactions excluded the old Williams School parcel, as well as two other small parcels at the north end of the Williams tract—a .63-acre lot that Daniel Williams sold to the Plymouth Railroad Company in 1869, and .725-acre lot that Daniel Williams sold to the Pennsylvania Railroad Company in March 1892.[11]

The reasons for this virtual gift of a sale from the Williams heirs indirectly to Widener are not clear. By virtue of Daniel Williams's will, following his death on June 21, 1899, Williams had left the entirety of this parcel to all of his heirs jointly. Their connection, if any, to Miller seems to have been tenuous at best. Miller, meanwhile, was the judge of the Court of Common Pleas in Montgomery County at the time of the transaction. Born in Weldon, Pennsylvania, to William and Caroline Miller on February 18, 1865, Miller lived the majority of his youth in Montgomery County. He was educated at a combination of local Friends and public schools, and he earned his law degree from the University of Pennsylvania in 1886. He practiced law in Philadelphia for six years while simultaneously serving as a partner in Miller Brothers, a wholesale tobacco dealer. In 1892 he moved his practice to Norristown, acting as a director of the Montgomery Trust Company for ten years and serving as the county solicitor from 1909 to 1914. Married into the Yeakle family of Whitemarsh, he was the father of four children—none of them married into the Williams line—and a staunch Republican. Indeed it may be that the only thing he had in common with the Williams heirs was membership in the Society of Friends.[12]

Widener appears to have used a similar methodology to acquire the old Coulston lands on the north side of upper Flourtown Road that same year. This parcel had been under James M. Coulston's control for many years, of course, but upon his death on March 24, 1901, the land was bequeathed to Coulston's widow and her heirs. Following Tacy Amanda Coulston's death in 1907, the property shifted to her children, most particularly Walter Coulston, who had been farming the plot since his father's death.[13] On December 30, 1916, John Donahue of Philadelphia acquired the 32 acres, sixty-nine perches of land from the old Coulston homestead. This plot represented the land on the north side of upper Flourtown Road that William Coulston had obtained from Isaac W. Potts in the early nineteenth century, including the original Isaac Williams House. How or when this land made its way into Widener's hands is not clear, but by 1927, this parcel was clearly part of the 600-plus acres under Widener's control, along with the old Hiram Corson parcel located just north of the Coulston parcel on the south side of Flourtown Road.[14] Albert Schmitz appears to have sold this latter, 23-acre parcel (the old Corson farmstead) to Widener on February 23, 1926, as part of a multiparcel transaction.[15]

Widener knew that rejuvenating the thoroughbred operations at Erdenheim would require individuals familiar with the old farm, so it should not seem surprising that he hired Dr. Anthony J. McCloskey, son of John McCloskey, who had for so long worked as Welch's chief assistant, Kittson's stable manager, and the Carsons' farm superintendent.[16] Widener also drew attention to Erdenheim through the renewal of regular races at the one-mile oval. Races there included

FACING: Daniel Williams House, north side of lower Flourtown Road on Sheep Tract. Photo by Eric Plaag, May 2012.

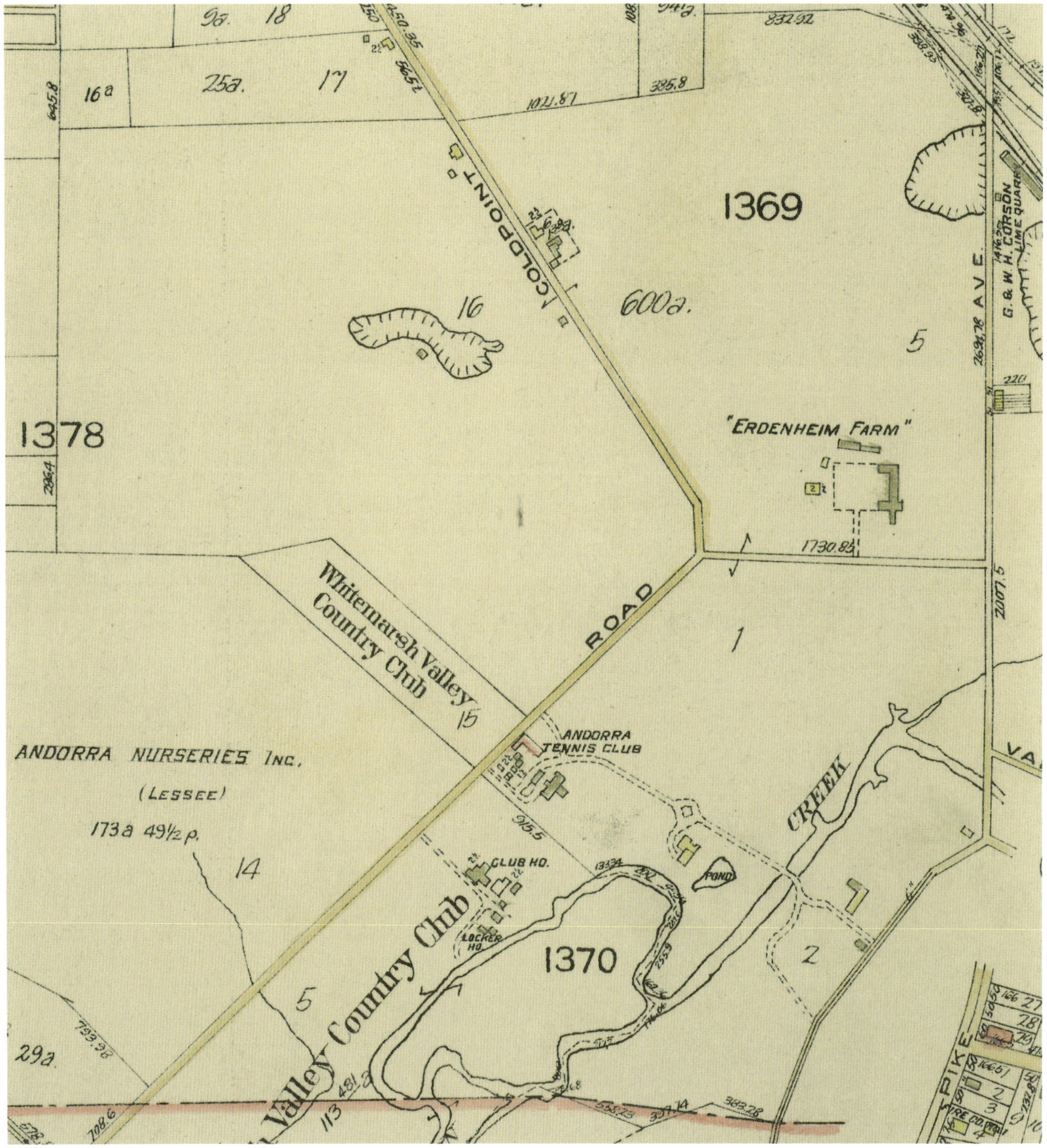

FACING: Dr. Hiram Corson House, upper Flourtown Road. Photo by Eric Plaag, September 2013.

Portion of Widener's Erdenheim lands west of Stenton Avenue as of 1927. Note that "Old" Erdenheim is no longer standing. Detail from *Atlas of the Reading Main Line, Montgomery County, Pennsylvania,* volume 1, Philadelphia: Frank H. M. Klinge, 1927, plate 32.

the Whitemarsh Valley Challenge Cup steeplechase race, which was first inaugurated at Erdenheim in 1916. Such events were a big draw for the society set, as evinced by the coverage of the 1920 race: "As the day was ideal the society maids and matrons all wore smart costumes, both sports togs and afternoon frocks. Parasols were also observed, which is rather unusual at an autumn race meet. All available parking spaces surrounding the racetrack were filled with automobile parties. Mr. and Mrs. Henry Brinton Coxe's automobile appeared to be the rendezvous for a number of the younger social set."[17]

But it was not only the wealthy who turned out for such events. The 1921 Whitemarsh Valley Fall Meet, for example, drew a crowd of more than four thousand for six separate races, one of which was a half-miler featuring horses owned by neighboring farmers whose lands were routinely used by the Whitemarsh Valley Hunt Club for hunts. The purse consisted of $50, with each entrant finishing under one and a half minutes receiving a horse blanket. Widener resumed the Erdenheim Cup, a two-and-a-half-mile race over open brush, that year, too, it not having been run since 1916.[18] While Widener continued a lengthy partnership with the Whitemarsh Valley Hunt Club over the next thirty years or so, by 1954, it was time for a change. That year the Whitemarsh Valley Hunt Club sold the 8.8-acre parcel on which the hunt club had sat for so many years back to Widener for $27,500.[19]

FACING: Former Whitemarsh Valley Hunt Club clubhouse, West Mill Road. Photo by Eric Plaag, September 2013.

Revolutionary Changes at "Old" and "New" Erdenheim

In addition to his efforts to rebuild Erdenheim's reputation as both a breeding establishment and racing destination, Widener also set to work almost immediately on transforming the buildings and grounds at Erdenheim. Initial targets appear to have been a collection of barns and other buildings located on the southern end of the Daniel Williams parcel, just above the bend in the Wissahickon, as well as a stone crusher building located near the Carson College boundary along Wissahickon Avenue. Widener followed up these demolition projects with the removal of a collection of homes at the southern extreme of Erdenheim on the northeast corner of Wissahickon Avenue and Stenton Avenue, properties that had existed at this location at least since Potts and Yeakle had a business concern there just prior to selling the plot to Welch.[20] Widener also appears to have demolished the existing barns at the Daniel Williams farm on the north side of Flourtown Road, replacing the barns quickly with new structures. Over the next thirty years, he made regular changes to this configuration until arriving by 1949 at something resembling the present arrangement of the barns at the Sheep Tract. Various ancillary structures behind the Isaac Williams House on the north side of upper Flourtown Road also disappeared between 1916 and 1938.

One of Widener's earliest projects must have been the demolition of many of the old Carson buildings at "New" Erdenheim, especially those that were close to the house. Nearly all of them were demolished before 1925, when a magnificent new greenhouse, several cottages, a garage, and an elaborate garden stood immediately to the west of their former location. Curiously, in spite of these

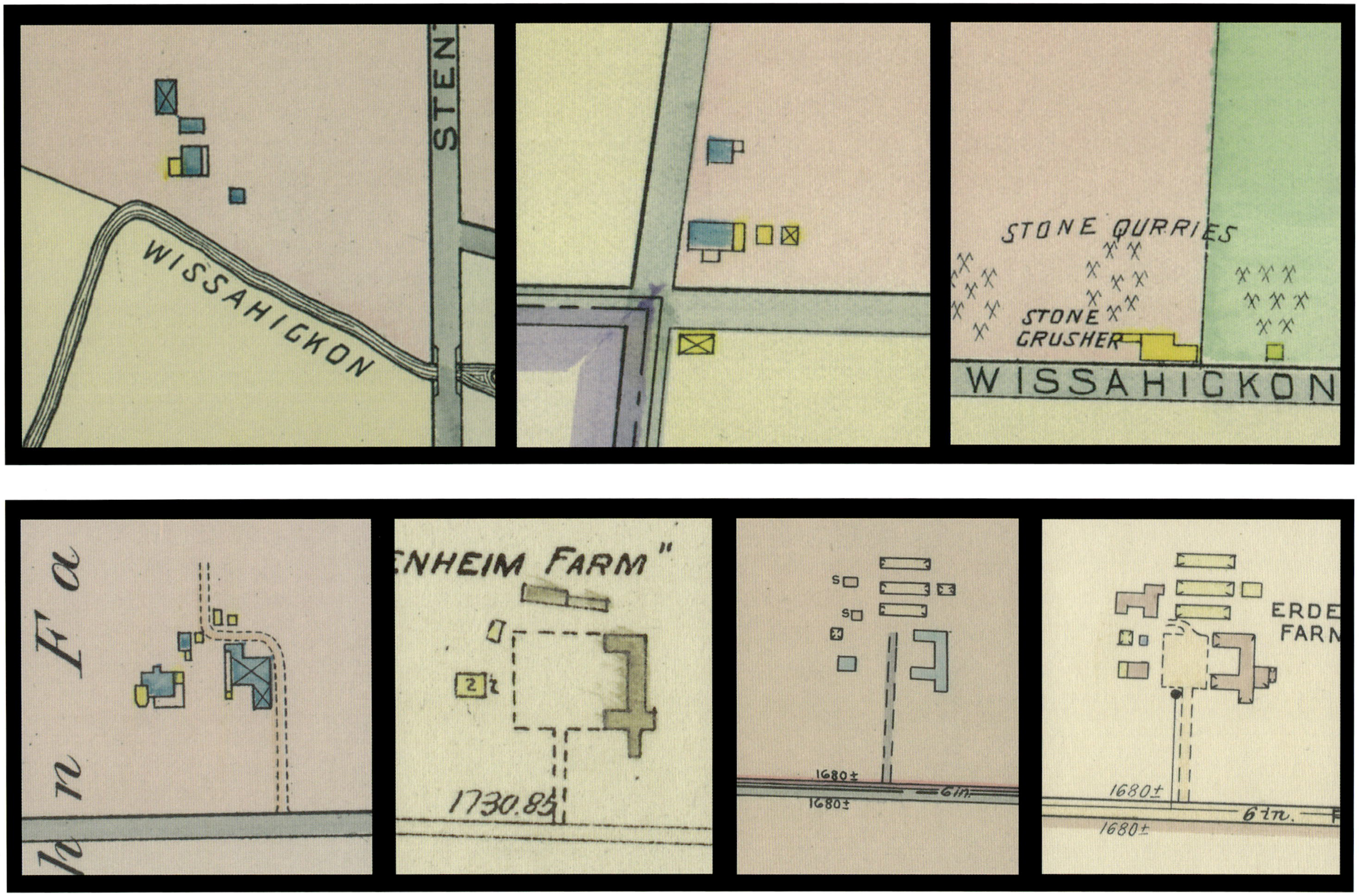

STEN
WISSAHICKON
STONE QURRIES
STONE
CRUSHER
WISSAHICKON
NHEIM FARM"
1730.85
1680±
6in.
1680±
1680±
6in.
1680±
ERDE
FARM

FACING, TOP: Williams houses on south side of Flourtown Road (left), houses at the intersection of Stenton Avenue and Wissahickon Avenue (center), and stone crusher among quarries adjacent to Carson College on Wissahickon Avenue (right). Details from plates 22 (left) and 20, *Atlas of the North Penn Section of Montomery County,* 1916, Map Collection, Free Library of Philadelphia.

FACING, BOTTOM: Changes at the Daniel Williams farm (Sheep Tract), 1916, 1927, 1938, 1949. Detail from *Atlas of the North Penn Section of Montgomery County,* 1916, plate 22; *Atlas of the Reading Main Line,* volume 1, 1927, plate 32; *Property Atlas of Montgomery County,* volume D, 1938, plate 20; *Property Atlas of Montgomery County,* volume D, 1949, plate 20. All images from Map Collection, Free Library of Philadelphia.

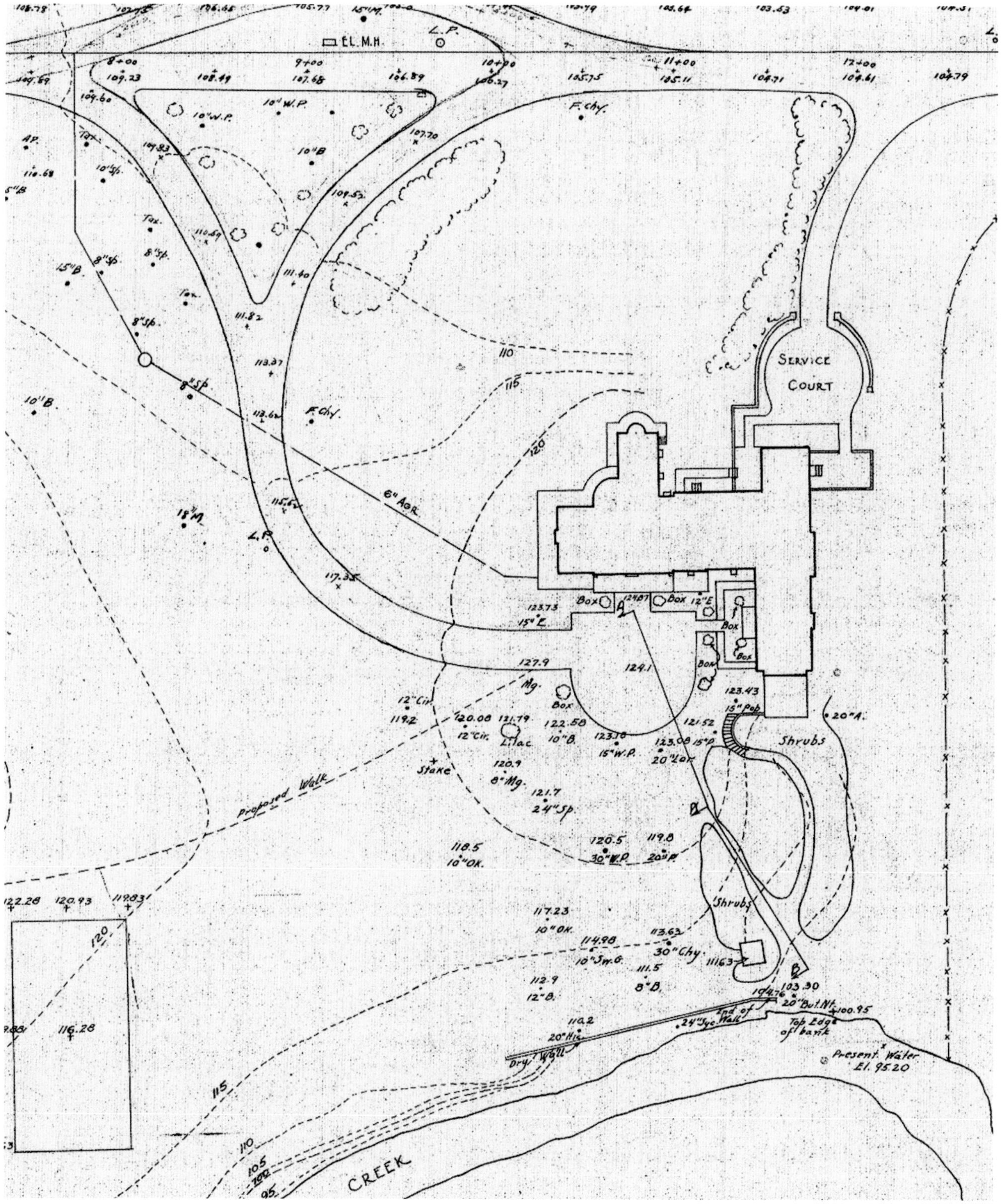

RIGHT: Detail from 1925 plot plan showing "New" Erdenheim, with existing tennis court to the southwest (lower left). The top of the Wissahickon oxbow is visible at bottom. Plan 28-Pt1, August 7, 1925, from Olmsted Associates Records, Frederick Law Olmsted National Historic Site.

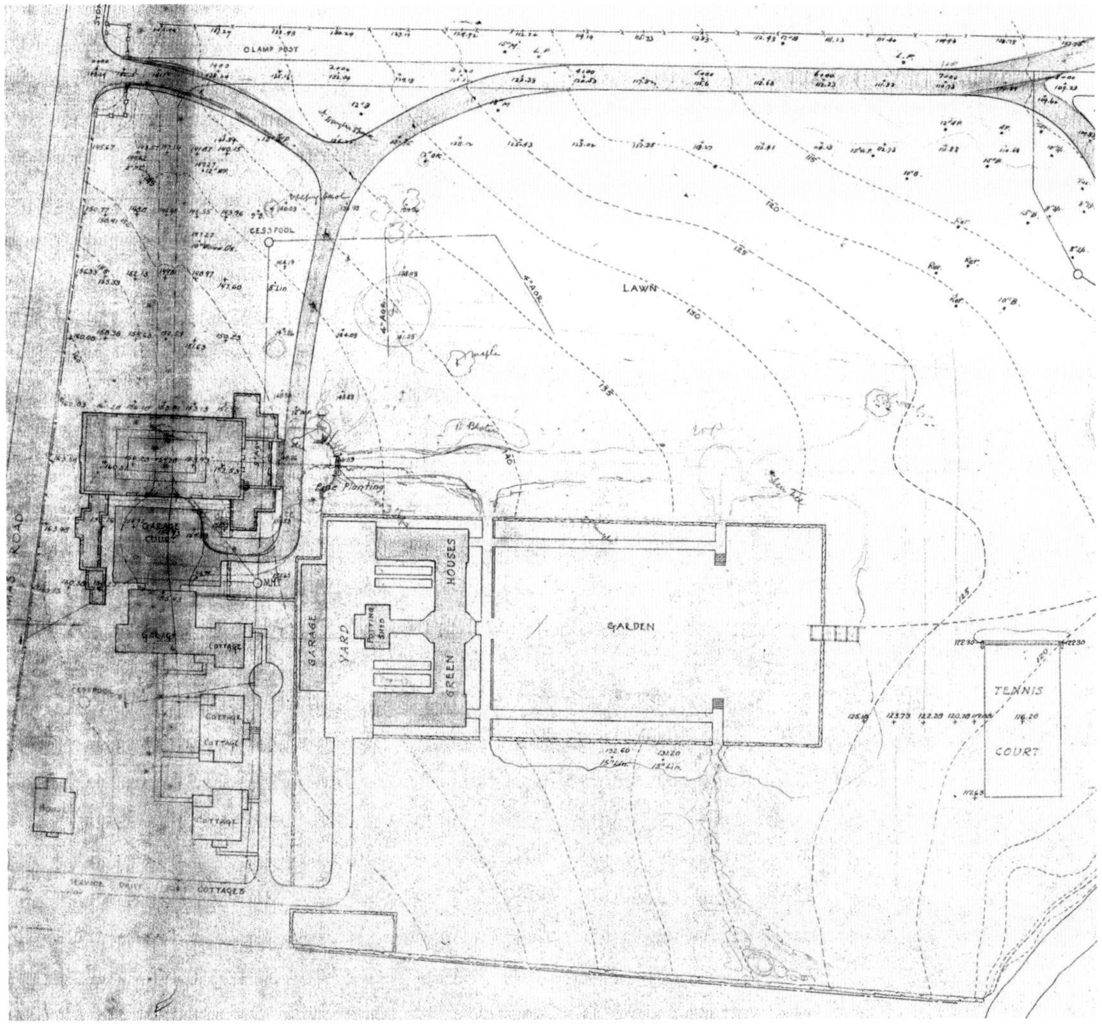

drastic changes to the west of "New" Erdenheim, Widener did retain the tenant house fronting on Thomas Road for several years, but by 1938 it was gone, too.[21]

Widener also dramatically altered the main house at "New" Erdenheim. Sometime between 1916 and 1920, Widener demolished the detached carriage house and outbuildings that lay immediately to the southeast of the main house, then erected an attached wing running perpendicular to the main house along the east side of the main house. Aerial images from this period suggest that Widener also filled in the terrapin pond to the east of "New" Erdenheim. Alterations to the main house also included a complete transformation of the exterior appearance of the building. The former one-story porch was removed and a colonial revival two-story porch—evocative of Mount Vernon's famous portico—took its place. Widener also transformed the two-story crenellated wing at the west end of the house to three stories with a dormer, and he either demolished the gambrel-roofed east wing or dramatically altered it, adding a dormered third story to a portion of it (to balance the revised west wing) and removing the chimney at the far east end of the east wing. Widener also added a dormered third story to the projecting wing on the north side of the house and added a complex service entrance at the northeast corner of the complex. Beyond these expansions, the overall effect of the renovations eliminated the medieval-style flourishes of Carson's version of the house and created the illusion that the

Detail from 1925 Plot Plan showing proposed location for new tennis club buildings, with existing garage, greenhouses, garden, and tennis court to the east. Thomas Road is to the west (left side of drawing). The main house ("New" Erdenheim) lay just to the east (right) of this portion of the grounds. The upper part of the Wissahickon oxbow is visible at lower right. Plan 28-Pt2, August 7, 1925, from Olmsted Associates Records, Frederick Law Olmsted National Historic Site.

property had long been a southern-style colonial plantation. As part of this renovation, Widener set a stone into the new chimney indicating that the house had been built in 1764—the same year so frequently quoted in early twentieth-century newspaper accounts as the date of construction of "Old" Erdenheim.[22] Given his obliteration of "Old" Erdenheim sometime in late 1924 or early 1925, one wonders whether Widener asked that this stone be moved from the old Hocker residence (which has so often been attributed to 1764) and installed in the Lukens/Kittson/Carson house chimney as part of Widener's renovations there to create a false sense of historic continuity between the parcels.

Widener Demolishes "Old" Erdenheim

Nothing survives to tell us why Widener elected to tear down the magnificent home at "Old" Erdenheim, in which Hocker, Welch, and so many other owners at Erdenheim had resided, but it most certainly was a choice he made as part of the numerous "improvements" he initiated at Erdenheim shortly after purchasing the property. While there appears to have been some hesitation initially (it was not slated for demolition on plans dated late in 1924), Widener completed "Old" Erdenheim's demolition no later than 1925. One reporter from *The Field Illustrated,* who visited Erdenheim for a March 1925 article, noted simply, "Many of the old stone buildings have been razed in the march of improvement inaugurated by Mr. Widener. One of these was the home of Mr. Welch for many years."[23] Some of these improvements were no doubt to suit the whims of his new bride, Jessie Sloane Dodge, the daughter of New York millionaire Henry T. Sloane.[24] Widener married her, as her second husband, on March 20, 1917, shortly after his purchase of the various Erdenheim parcels. Widener typically made these alterations when he and Mrs. Widener were not at Erdenheim, however; given their ownership of a plantation at Mackay Point, South Carolina (immediately south of Pocotaligo), where the Wideners were reported to spend six to eight weeks each year, from late November to March, facilitating changes at Erdenheim without Mrs. Widener present must have been relatively easy. While Widener hunted quail and duck in the swamps of South Carolina, work continued apace at Erdenheim.[25]

Demolishing the original Erdenheim residence and dozens of other buildings on the surrounding Erdenheim parcels was not the only change Widener brought to the old, historic landscape. Nearly two centuries after the first headstone was erected within its borders, the old burial ground once contained within the boundaries of the Lukens parcel still stood in 1915 as a sentinel on the slope in front of "New" Erdenheim. As one local commentator indicated in a 1904 column, "It is a beautiful site for the repose of the dead. . . . A short, but steep, declivity, separates the graveyard field from the vale, near the Wissahickon."[26] Whatever its appeal to local historians, though, upon transfer of title from the Carson estate to the Wideners, the latter apparently arranged for the complete demolition of all above-ground evidence of the old burial ground. William R. Yeakle, an eyewitness to those events, described what occurred when his father discovered that the cemetery had been obliterated from the property:

> Upon being questioned on the matter, the new owner informed [Yeakle's father] that he knew nothing about it and that he supposed the executors of the Carson estate might know, if any one did. A representative of the Carson estate was next seen and he declared that the new purchaser had stipulated, in the conditions of sale, that the old grave-yard must be destroyed. This same representative also said that he did not know what had become of the walls or the grave-stones but that possibly the contractor, who did

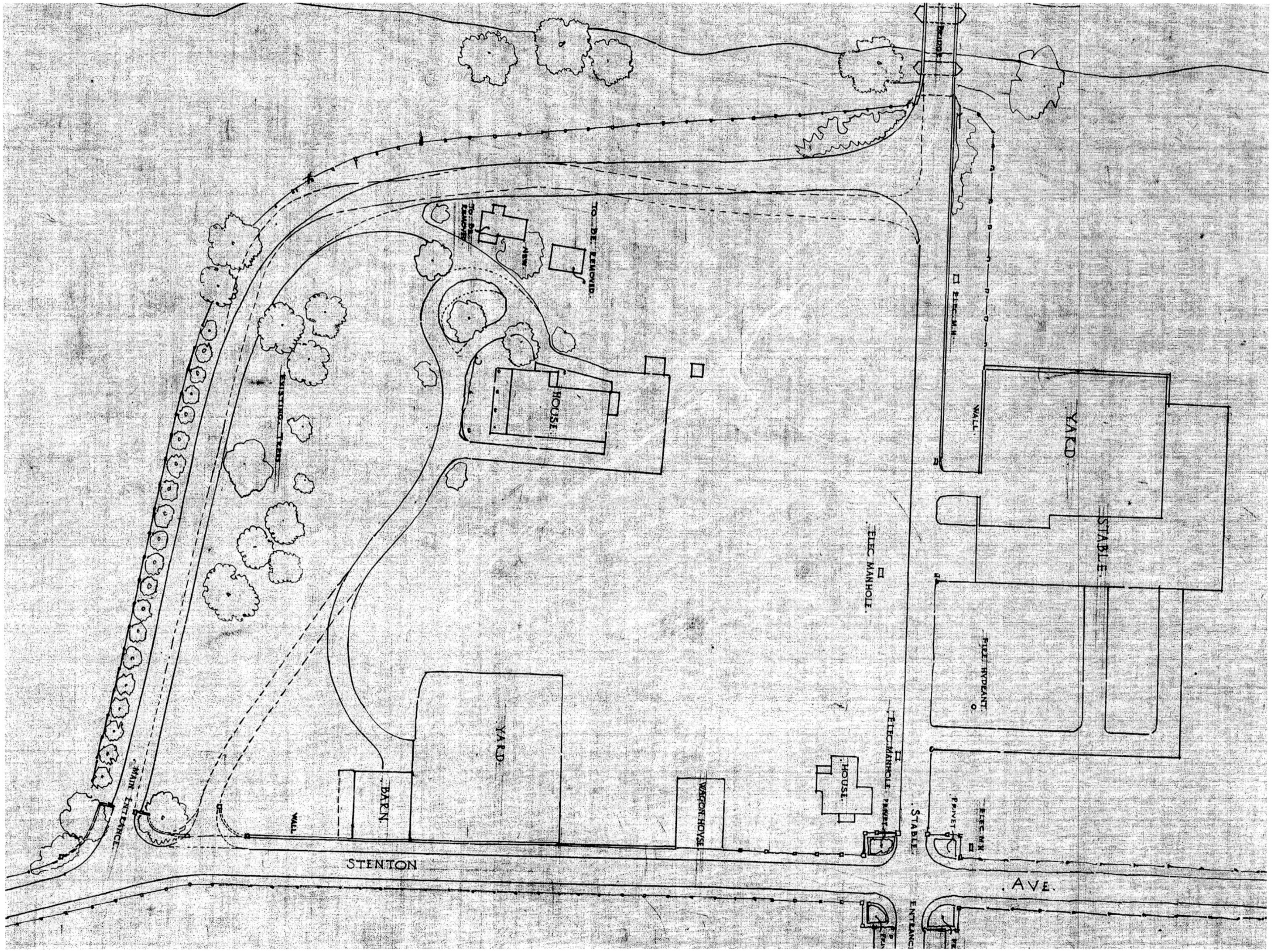
STENTON
AVE.
MAIN ENTRANCE
EXISTING TREES
TO BE REMOVED
HOUSE
YARD
BARN
WALL
WAGON HOUSE
HOUSE
ELEC. MANHOLE
ELEC. MANHOLE
STABLE ENTRANCE
FIRE HYDRANT
WALL
YARD
STABLE
PRIVET

> the work, might know. The contractor was next questioned and replied that he did not know where the stones were and that, in all probability, the workmen had broken them and thrown them away. Those were built for permanence and were not easily demolished and necessitated speedy and laborious action to accomplish the destruction almost "over night." This briefly is the story as far as we know it.[27]

Soon after his purchase of the Carson Estate lands, Widener also set his sights on the old Williams School parcel, which sat directly to the southeast of the intersection of Flourtown and Thomas Roads, squarely in the midst of the land Widener had just obtained. Knowing that the one-story school building had recently closed as part of a consolidation move within the township, Widener approached the Whitemarsh Township School Board and offered $5,000 for the building and land, an offer the board quickly accepted. Widener demolished this property after the purchase as well.[28]

Not all of Widener's alterations at Erdenheim involved demolition, however. Sometime between 1916 and 1927, Widener constructed a new residential dwelling on the old Welch track site just east of Stenton Avenue and south of West Mill Road (by then also called Flourtown Road). It was here that Andrew

FACING: Survey detail of "Old" Erdenheim, from *General Plan of Estate of George D. Widener, Esq.,* with revised layout of roads and garden planting, February 2, 1918, surveyed December 12, 1917, traced November 2, 1920, sheet dated November 7, 1924, Olmsted Associates Records, Frederick Law Olmsted National Historic Site. This plan includes proposed changes to the roadways at "Old" Erdenheim, as well as numerous changes to the gardens at "New" Erdenheim. Note that "Old" Erdenheim is still standing, although nearby buildings are slated for demolition. "Old" Erdenheim would be demolished by 1925.

Aerial view of "New" Erdenheim as it appeared circa 1924, following house renovations but prior to landscape renovations by George D. Widener Jr. Note that the pond to the east of the house is gone here, but the bridge over the Wissahickon is still an iron truss bridge. Courtesy Springfield Township Historical Society.

Jackson Joyner, Widener's principal horse trainer during the 1920s and 1930s, resided.[29] Just to the south of the driveway for this residence (which aligned with the driveway leading to the Wissahickon Bridge on the west side of Stenton), Widener also built a massive stable and exercise building, replete with thirty stalls—a building that still stands today on the property.[30] Meanwhile, just to the west of the former stone crusher site and the quarries along Wissahickon Avenue, Widener built a large residential property that appears to have served originally as a dormitory or apartment house for employees at the farm. It remains on the property.

About this same time, in 1920, Widener also opened up a portion of his property to public traffic by selling two rights of way to the Commonwealth of Pennsylvania, each to be a continuation of Fairmount Park and allow riders to conduct their horses over his land. One of these passages, a narrow buffer strip between the road and the equestrian farm totaling just over 8 acres, led from the southeast side of the intersection of West Mill Road (Flourtown Road) and Stenton Avenue south to the Whitemarsh-Springfield line. The other was a triangular piece of land, totaling just under a half acre, on the west side of Stenton immediately opposite the terminus of the first right way at the township line.[31]

As previously mentioned, Widener also engaged in extensive renovations of the old Lukens tract, virtually obliterating many of the buildings and features that had been positioned to the west of the house approaching Thomas Road. Within only eight years, however, Widener seemed dissatisfied and called for additional alterations to the west end of this parcel. To accomplish these changes, in 1924 the Wideners hired Olmsted Brothers, the legacy firm of Frederick Law

FACING: Circa 1920 Widener stable and exercise building, east side of Stenton Avenue. Photo by Eric Plaag, September 2013.

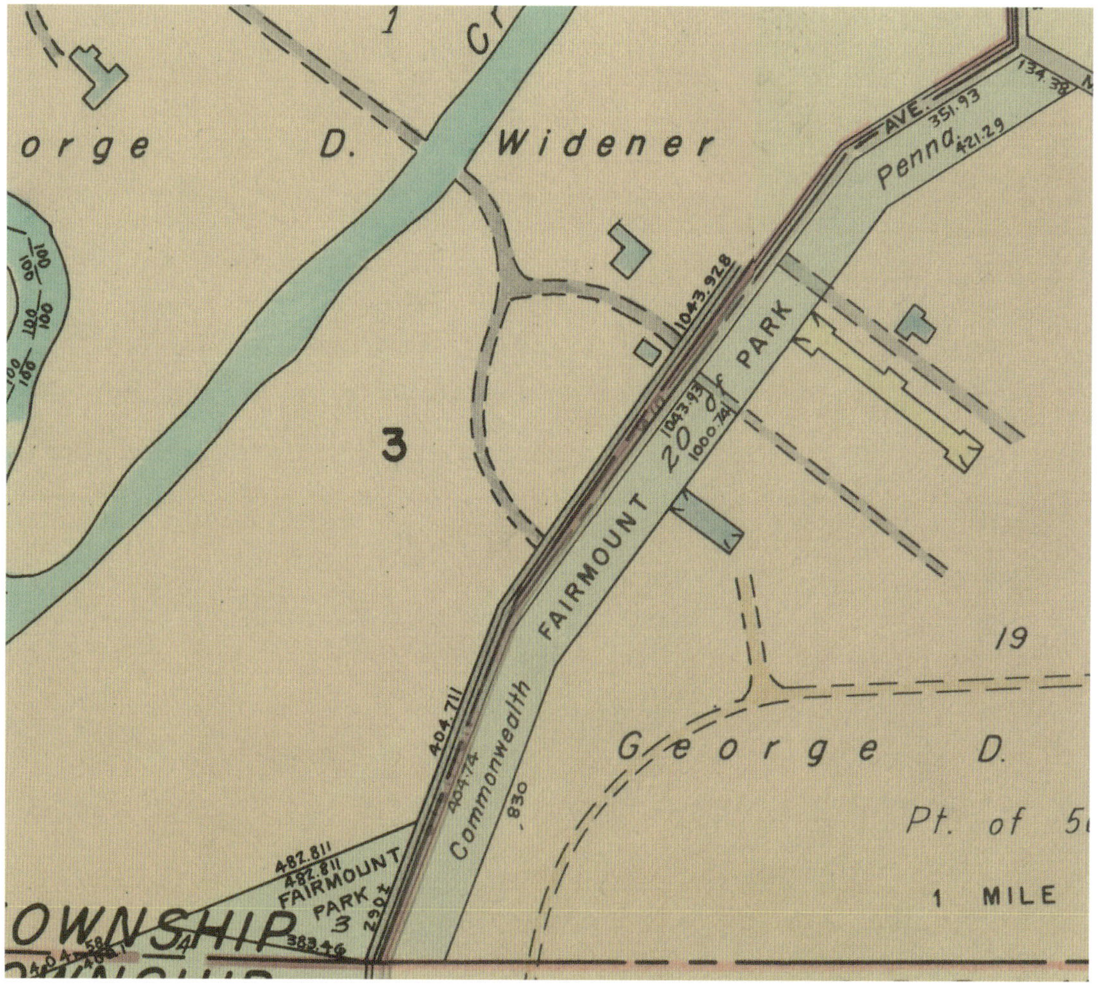

Composite image of Widener parcels transferred to Commonwealth of Pennsylvania for Fairmount Park, 1920, detail from plates 20 and 17, *Property Atlas of Montgomery County, Pa.*, volume D, Philadelphia: Franklin Survey Company, 1938, Map Collection, Free Library of Philadelphia.

Williams
READING
R.R
AHICKON
AVE

Olmsted's old landscape architecture business, Olmsted Associates, now run by his sons. These changes were completed in a series of projects, beginning with a complete revision of the gardens and drives at "New" Erdenheim for nearly $40,000.[32] By 1925 Widener was also planning a new indoor pool and tennis house for these grounds—apparently using Horace Trumbauer, the architect of so many other Elkins and Widener properties in years past, including Widener's childhood home—and requested an estimate from Olmsted Brothers for the entrance court and garden for this addition, which was to replace the old Carson tennis court house located closer to the main dwelling.[33] Having a tennis club necessitated having a new clay outdoor tennis court, too, and by May 1925, plans were underway for this addition.[34] Maps from 1927, after the tennis clubhouse building was complete, suggest that Widener had named the new facility the Andorra Tennis Club—no doubt a nod to the Andorra Nurseries located just yards south from the new tennis club on Thomas Road.

In addition Widener requested new plantings to screen out the appearance of the White Marsh Valley Country Club to the south of "New" Erdenheim and the view from the creek toward Thomas Road, as well as the removal of a lilac bush, an old poplar tree close to the house, and other shrubs that were blocking a clear view of the creek.[35] One of the most challenging requests from Widener, however, was his desire to rework the roadways on the east side of

FACING: Aerial view of Erdenheim, 1928, detail from *Aerial Survey of the Philadelphia Region,* Regional Planning Federation of the Philadelphia Tri-State District, circa 1928, plate 50, Map Collection, Free Library of Philadelphia.

Main House at "new" Erdenheim, as it appeared in 1930. PH92, Olmsted Associates Records, Frederick Law Olmsted National Historic Site.

Renovated gardens at "New" Erdenheim, looking south, 1930, PH103, Olmsted Associates Records, Frederick Law Olmsted National Historic Site.

the Wissahickon, near the site of "Old" Erdenheim, culminating in a large entrance gate at Stenton Avenue. Initial plans suggested that the straightest path would take such a road through "the horse graveyard" where Leamington, Flora Temple, and Maggie B. B. were buried, but later plans allowed for a curve to circumvent this problem. At the same time, the long, straight driveway from Stenton Avenue to the bridge across the Wissahickon was severed, and though it is not explicitly mentioned in the plans, it appears that Widener replaced the old iron truss bridge over the Wissahickon with the current stone one during this time.[36] All these changes appear to have been requested by Widener to eliminate the tendency of some drivers to use his driveway and bridge across the Wissahickon as a shortcut between Stenton Avenue and Thomas Road.[37] Other alterations called for a new rose garden, 100' by 120', as well as new arbors flanking the existing cut flower garden areas.[38]

For decades, many keen observers had suggested that the Wissahickon's waters were critical to the success of the horses produced and raised at Erdenheim. As one commentator noted in 1935, "The creek bares the limestone formation which, with the excellent bluegrass, is largely instrumental in deciding the quality of horses raised on the farm."[39] As it turned out, though, that lime in the soil that was so good for the health of Widener's horses was not particularly kind to the rhododendrons and azaleas, and this required replacing much of the soil in the garden beds to keep the plants from dying. Solving the problem of the lime in the water also required the addition of a connected tank of hydrochloric acid. This, too, created its own problems, because the chemical effect removed the lime from the water but created a substantial deposit in the tank.[40] It was, in short, yet another instance of Erdenheim's historic hallmarks getting in the way of Widener's plans for his farm.

Renovated gardens at "New" Erdenheim, looking west, 1930, PH 106, Olmsted Associates Records, Frederick Law Olmsted National Historic Site.

RIGHT: Gardens at "New" Erdenheim with new trellis at east end, 1926, looking west. PH37, Olmsted Associates Records, Frederick Law Olmsted National Historic Site.

More Fires at Erdenheim

Fires continued to be a problem at Erdenheim under Widener's ownership of the property. One of the most devastating of these was a 1921 blaze that destroyed the $40,000 stable building located on the northeast side of the driveway connecting Stenton Avenue with "New" Erdenheim, as well as $15,000 of farm implements and equipment. The fire was discovered just before 7:00 P.M. on the night of January 16, when William Tomlinson, the farm's watchman, "saw the flames spurting from the second story of the stables, a frame building two hundred feet long and one hundred feet wide, with an extended wing in which the prize cattle were stabled."[41] Tomlinson then cried out to Edward Stanton, Erdenheim's farm manager, who occupied the house fronting on Stenton Avenue, immediately opposite the main horse stables. Through their joint efforts, Erdenheim employees rescued the prize cattle, about twenty horses, and dozens of sheep. Only one ewe was lost. Once the firemen and apparatus finally arrived from a dozen surrounding communities, high winds forced them to direct their attentions to the nearby residences and horse stables, for fear that they might be lost as well. The cause of the fire was not discovered.[42] While there is no direct evidence to suggest that damage from this event may have led to "Old" Erdenheim's demolition, the only residences immediately adjacent to this stable would have been the

New barn at "Old" Erdenheim, 1926, with Stenton Avenue superintendent house and Welch stables in the distance, following diversion of road. Note that bridge over the Wissahickon is no longer an iron truss. PH85, Olmsted Associates Records, Frederick Law Olmsted National Historic Site.

"Old" Erdenheim house originally built by Hocker, the old Leamington stable converted into an office by Carson, and the superintendent's house fronting on Stenton (which still stands).

Widener's Champions

With the many changes taking place at Erdenheim, it may seem surprising that Widener displayed lightning speed and efficiency in restoring Erdenheim's reputation as a both a breeding farm and a training facility for America's best racehorses. All told, Widener's successes at the track are comparatively staggering. Over the arc of his career, horses raced by Widener won more than 1,240 matches, including the 1962 Belmont Stakes, five Eclipse Awards, and five Travers Stakes, and raised more than $9,000,000 in purse receipts for Widener; the horses he and his wife bred after 1922 earned more than $16,000,000 for their owners. He was especially successful with Futurity Stakes winners, including such noteworthy horses as St. James, Jamestown, Battlefield, Jester, and Bold Hour. Nevertheless Widener was adamant about avoiding the Kentucky Derby, insisting that a race of that length in May was too hard on three year olds.[43]

Many found his position absurd. One columnist, writing after Widener had refused to enter Jaipur into the Kentucky Derby as a three year old, observed, "One of the most farsighted men I know, George D. Widener, . . . can see four months ahead. He knows his colt will be ready for the Preakness on May 19, but won't be ready for the Derby two weeks earlier." Noting that Widener had been quoted in *Sports Illustrated* as saying that the Derby, with horses running ten furlongs at an average age of thirty-seven months, was "too far, too early," the columnist piled on: "This is a tired old argument that stopped making sense over thirty years ago. It is demonstrably false. . . . To hear Widener and all the other

George D. Widener Jr. (far right), at the Belmont Park Jockey Club Gold Cup Presentation, October 5, 1946. Jockey Eddie Arcaro is second from the left and Walter M. Jeffords at center. The other two men are unidentified. Cook Collection 445, Keeneland Library.

"St. James, McTaggart Up," from Hervey, *Racing in America, 1922–1936*, Keeneland Library.

Eastern mossbacks tell it, the Belmont is the perfect race at the perfect time. And yet, if you look at the calendar, [twelve furlongs] is cruel and unusual punishment for colts thirty-eight months old."[44] Widener must have been doing something right, though, since Jaipur—perhaps Widener's most famous horse—won the Belmont that year.

From the beginning of his ownership of Erdenheim, it was clear that Widener was serious about both breeding and racing. He won his first race in 1916, when Columbine crossed first in the Walden Stakes at Pimlico.[45] In 1917, just months after acquiring Erdenheim, Widener announced his intention to place sixteen horses in the Futurity for 1919, and he already had plans to race several British and American two and three year olds in the 1917 season.[46] Widener's service stateside as a lieutenant in the Quartermaster Corps at the end of World War I does not appear to have significantly slowed his progress as a horseman, and in 1923 he expanded his breeding operations with the purchase of the Old Kenney Farm, a subdivided parcel of the historic Elmendorf Stud (Widener's uncle Joseph Widener, and later his cousin Peter A. B. Widener II owned the remainder of Elmendorf until 1950). Widener thereafter split his breeding operations between Erdenheim and Old Kenney Farm until 1968, when he sold the latter operation.[47] During this period of rapid expansion and remarkable success, Widener also purchased St. James in 1922, a yearling "of fine size and an extremely aristocratic stamp, with a head and neck of pictorial outlines," for $9,000. While his racing days were cut short by injury, he won four of his seven starts and placed second in two other races, earning $94,035—a record for an American horse with such a brief career.[48]

Such a scale of success was commonplace among Widener's horses. In 1930 alone Widener's Jamestown scored $151,925 in earnings after winning the Flash, United States Hotel, Saratoga Special, Grand Union, and Futurity Stakes; indeed, to that point, no other two year old had made a sweep of the Saratoga races. Yet in spite of these accomplishments with Jamestown, Widener maintained his usual tradition and withheld Jamestown from the Kentucky Derby and the Preakness in 1931 on the advice of his trainer, Andrew Jackson Joyner.[49] Jack High was another of his champions during this period, beating a field of fifteen in the 1928 Hopeful Stakes by a length and a half. He won three of the five races he ran in 1929, and four of the six he ran in 1930, winning the Metropolitan that year in an American record time for a mile of 1:35. In 1931 Jack High won two of his three races and in 1932, three of his ten while simultaneously standing stud at Erdenheim.[50]

In 1930 Widener also had phenomenal success with Jamestown, a horse foaled at Erdenheim in 1928 and trained by Joyner, when Jamestown brought home a staggering $151,925 that single year—a sure record had it not been surpassed by his best competitor that year, the immortal Equipoise. Following that season, many called Jamestown the best two year old since Man o' War. While Jamestown enjoyed scattered successes thereafter and helped force Equipoise to set a new American mile record, he never replicated his level of accomplishment as a two year old and was soon sent to Widener's stud in Kentucky.[51] In 1936 Widener's three-year-old filly High Fleet followed up on her previous year's performance of winning, placing, or showing in every race she started by winning eight times and placing second in the other five of the thirteen races she ran.[52] His horse Eight Thirty, bred by Widener and winner of the Travers in 1939, had several years of successful racing before retiring to stud and performing well for

George D. Widener's racing silks, based on contemporaneous descriptions and Frederick M. Burlew's *American Racing Colors: 1915–1930*, volume 3. Design by Kevin Archie.

GEORGE D. WIDENER

FACING: Jamestown, 1930, from *Turf and Sporting Digest,* Keeneland Library. Based on a similar image in Hervey, the jockey appears to be L. McAtee.

Eight Thirty, D. Richards up, Toboggan Handicap, Belmont Park, May 13, 1940. Morgan Collection 94953, Keeneland Library.

7

Widener in the following years. Eight Thirty's death on April 7, 1965, at the age of twenty-nine gave Widener a taste of the bittersweet, however. That afternoon Widener's filly What a Treat won the Prioress Stakes at Aqueduct.[53]

Finally in 1944 Widener seemed ready to break with his long-standing tradition of keeping his ponies out of the early spring races like the Kentucky Derby. That year he submitted the names of Platter and Lucky Draw to run in the Derby, and both were selected. Nevertheless Widener never let them run.[54] Asked about this controversial decision years later, trainer Bert Mulholland explained, "It has got to be such a circus that nobody like us cares about it too much. It's too early in the year and rips the horses apart. It's a race that really is the survival of the fittest."[55] In 1946 Widener generated more controversy when he refused to allow Lucky Draw to enter the Pimlico Special because of a new track requirement that horses be locked away in a special receiving barn for two hours prior to any race—a policy he regarded as undermining the reputation of the racing world.[56] While Widener always seemed to have a champion or two dominating the headlines in any given year, other noteworthy horses during this later period of his career included Battlefield (winner of the 1950 Futurity Stakes and Saratoga Special), Lights Up (winner of the 1950 Travers Stakes), and Bold Hour (winner of the 1966 Hopeful Stakes). To put Battlefield's value to Widener in perspective, Bert Mulholland purchased Battlefield on behalf of Erdenheim for $4,500, then the horse promptly set a record among two-year-old colts with winnings of $198,677.50.[57] Among all of them, though, one of Widener's favorites was Ambergris, in large part because Widener thought he possessed the most beautiful thoroughbred head he had ever seen.[58]

FACING: Battlefield wins the 1950 Belmont Futurity Stakes by a nose. Morgan Collection A70092, Keeneland Library.

The champions bred at Erdenheim were not just George Widener's. Widener's wife, Jessie Sloane Widener, formally entered the racing scene in 1930 and also bred horses at Erdenheim, including six Stakes winners and the champion Evening Out.[59] She was regularly present at many of the races sponsored by Widener's interests or in which their horses were running, and she was on hand in 1967, when a Widener horse won the Widener Handicap for the first time. Mrs. Widener was also a champion in other pursuits besides horseracing. A frequent entrant at the Philadelphia Flower Show, Mrs. Widener—along with occasional contributions by her husband—cultivated blooms in the elaborate gardens at "New" Erdenheim and won numerous awards as a result. She died at the Miami Heart Institute on March 11, 1968, at the age of eighty-four.[60]

In 1942 the Westchester Racing Association, which operated Belmont Park, elected Widener to be its chairman, a position he held until 1955. Perhaps his greatest legacy from this office was his introduction of a grass course at Belmont, later named the Widener Turf Course in honor of Joseph E. Widener.[61] As the years went on, Widener became widely regarded as a veritable god of the turf. In 1950 the Jockey Club elected Widener as its chairman, a position he held for the next fourteen years. When he retired from that post in 1964, the Jockey Club would have none of it, electing him as their honorary chairman.[62] Following Widener's 1962 sale of the Old Kenney Farm in Lexington to Leslie Combs II (who renamed the property "Spendthrift Farm"), Widener continued to keep his breeding stock there and shifted the entirety of his breeding operations to Spendthrift. Erdenheim, meanwhile, remained his primary breaking and training facility.[63]

Widener, Gambling, and Baseball

One of the more intriguing moments in the career of Judge Kennesaw Mountain Landis, the longtime commissioner of Major League Baseball, came in the fall of 1943, when news leaked that Widener was a major stockholder in the Philadelphia Phillies. Ever since the 1919 Black Sox Scandal, when members of the Chicago White Sox were accused of throwing the World Series as part of a gambling scheme and Landis was forced as the new commissioner to clean up the public relations mess, Landis had worked hard to ferret out and ban gamblers and those who in any meaningful way endorsed or supported gambling. He also insisted that they be prohibited from having any ownership stake in major league clubs. Those in racing circles were Landis's favorite targets, but even such seemingly innocuous celebrities as Bing Crosby had been stripped of their ownership interests or blocked from making major investments in the clubs.

Widener's reaction to the exposé was swift and firm. When one intrepid reporter contacted him at Erdenheim and demanded to know what he'd do if Landis tried to force him out, Widener was clear—Landis would have to take the matter to court, because Widener had no intention of selling.[64] Indeed Widener was not new to ownership in other sports concerns, having had a significant stake in the ownership group of the Philadelphia Arrows, the city's hockey team, from 1929 to 1935, and he was probably confident that the law was not on Landis's side.[65]

FACING: Jaipur coming down the stretch, with Eddie Arcaro up, at Belmont's Cowdin Stakes, October 2, 1961. Morgan Collection T1980, Keeneland Library.

Had Landis taken Widener to task, it probably would have been a serious public relations miscalculation. Over the course of his career, Widener established a reputation of being one of the finer, most ethical, and more respected horsemen in the profession, in large part because of his opposition to night racing and off-track betting, as well as his vehement objection to the use of painkillers in horses to improve performance.[66] Similarly he was a vocal opponent of stakes purse inflation by tracks competing with one another, and he instead urged higher purses for less significant races as a means for controlling this problem—competing fields could still make their money without driving each other into bankruptcy or privileging already wealthy owners over those just getting their start in the business.[67] Widener and his uncle, Joseph Widener, also established reputations for themselves early on in the career of George Jr. when together they took on a costly legal fight to demonstrate the legitimacy of horseracing as a business rather than an expensive hobby. Confronted by the Internal Revenue Service with a bill for $403,000 in back taxes for losses the IRS said the Wideners inappropriately deducted between 1919 and 1922, the Wideners appealed their case to the U.S. Board of Tax Appeals and, after five years of arduous legal wrangling, won. The victory was a boon for the Wideners and dozens of other large-scale breeders and trainers.[68]

Andrew Jackson Joyner and W. F. (Bert) Mulholland

Born Andrew Joyner in Enfield, North Carolina, on August 18, 1860, and the son of Dr. Henry Joyner, who intended for his son to grow up to be a physician, Andrew Jackson Joyner served for decades as the trainer to Widener's horses at Erdenheim. After growing up in Halifax County, North Carolina, Joyner eventually found work as a postmaster at Weldon, North Carolina. The story ran that one day in 1878, Joyner was rummaging through the undeliverable mail, ripped open a package, and found dozens of images of the most popular racehorses of the day. From that point forward, he was fascinated with horses and endeavored to do everything possible to be a part of the horseracing circuit.[69] After securing employment as a stable boy in 1879, Joyner went on to train for a half dozen owners, including James B. Haggin, Sydney Paget, August Belmont Sr., David Gideon, W. C. Whitney, and ultimately Widener.[70]

Renowned for a "screeching whoop" that some equated to a rebel yell when he encouraged his horses, he was regarded as unrefined by some of the stuffy attendees at Ascot and other courses, but the owners he worked for knew better. After running his own stable for five years, from 1890 to 1895, where he accepted horses belonging to other owners, his reputation was sufficient that the owners began to hire him onto their farms. In 1908 he was regarded as America's most successful trainer, handling the winners of seventy-one races and generating over $170,000 in winnings that year alone. After seven years of working for owners in England (through 1915), he returned to America and joined the staff at Erdenheim, where he worked for the rest of his career, shepherding such champions as St. James and Jamestown. He also assisted with operations at Old Kenney Farm. As age and infirmity caught up with him in the 1930s, he relinquished many of his duties to Widener's new assistant trainer, W. F. (Bert) Mulholland. Even after his general retirement from racing, though, Joyner continued to live at Erdenheim, well into his eighties, still advising Widener on the best treatment and preparation for some of his horses.[71] Remembering Erdenheim late in his career, Joyner saw the training track (the one-mile oval) as critical: "It is one of the best training tracks in the country, and to it I attribute much of the success I have had with Mr. Widener's horses in the past twenty-five years."[72] So beloved by Widener was Joyner that after Joyner's death at Erdenheim in the summer of 1943, Widener commissioned Joyner's friend Neil Newman to write a serialized account of Joyner's life that ran for more than a year in *Blood-Horse* magazine.[73]

Mulholland, meanwhile, spent much of his early career living in Joyner's considerable shadow, to the point that some commentators insisted that a retired but still present Joyner was the puppeteer behind Mulholland's perceived successes.[74] Born in Lexington, Kentucky, on August 27, 1883, Winbert F. Mulholland—like Joyner—entered the racing world in 1899 as an exercise rider and stable boy for his uncle at the age of sixteen, served briefly as a jockey, then quit the racing world to become a movie house projectionist in New York. After several years Mulholland realized how unsatisfied he was and tried to make his way back onto the turf, securing a position as Joyner's assistant at Erdenheim in 1923. As Joyner's health failed in the 1930s, Mulholland took over his mentor's post at Erdenheim in 1933. Among the champions he trained were Eight Thirty and Lucky Draw (whom Mulholland considered to be the best horses he ever worked with), Jaipur, Battlefield, Lights Up, Evening Out, and Bold Hour. In an April 1964 interview, Mulholland chalked up his forty years of success working for Widener to the fact that the two men "never had a serious disagreement in all that time." In August 1967 the eighty-three-year-old Mulholland found his own health failing, though, especially following a "severe coronary occlusion" that occurred while he was on the job at Erdenheim. Despite being elected into the National Racing Hall of Fame at Saratoga shortly after his attack, Mulholland never recovered and could not return to his duties. He died July 11, 1968, at Germantown.[75] In the years following Mulholland's death, Syl Vietch served as the trainer at Erdenheim.[76]

A Man of Many Hats

In addition to his fine reputation as a horseman and breeder of champion thoroughbreds, Widener also parlayed Erdenheim's decades of success in breeding prized sheep and cattle into national fame. As early as 1920, for example, Widener's advertisements in international publications highlighted Erdenheim's successes. Describing the champion Draconis Duke, for example, one ad touted the herd sire as "a real show bull and a producer from a family which has made good." The claim, it seems, was not without merit, given that Draconis had already "won more prizes than any other cow of the Jersey breed," including "seven first prizes on the Island, winning first over a class of 37 and 78 entries, respectively," two Silver Rose Bowls, four championships, four sweepstakes, and the Theatre Challenge Cup. In this same advertisement, Widener also heralded Erdenheim's prize Cheviot sheep, including "Erdenheim's Royal Winner No. 13000, the only Royal Winner Cheviot ever brought to this country."[77] Just a year earlier, just one Jersey cow, Fern's Gamboge Linda, brought $850 for Erdenheim Farm at the annual auction by T. S. Cooper and Sons in Allentown, Pennsylvania, while Gypsy Maid of St. Mary's brought $775. Brampton Prince's Agatha, a heifer, brought another $700, while Rowers Milkmaid, an imported cow, earned $1,050 for Erdenheim. Such figures were common.[78]

Erdenheim's fame for such high-quality livestock reached international levels. In October 1968 Widener invited nearly one hundred members of the Philadelphia Society for Promoting Agriculture and their guests to Erdenheim for a tour of the facilities. Guests marveled as Bill the sheep dog, under the guidance

FACING: W. F. Mulholland, George D. Widener, Hedley J. Woodhouse, and Mrs. George D. Widener, May 26, 1954, Belmont Park, National Stallion Stakes, Morgan Collection A87533, Keeneland Library.

of Widener's shepherd Henry Kuykendall, kept watchful eye over the Cheviots in the field below "New" Erdenheim, while others were pleased for the chance to see Kanval, the champion ram of three Australian shows. Here guests learned that Widener would be sending several sheep to Toronto's Royal Winter Agricultural Show the following month. Stanley R. Gates, Erdenheim's superintendent, then toured guests out to the more distant reaches of the farm, allowing them to see some of Widener's 180 Hereford cattle and chat with trainer Ralph Delaney at the thoroughbred stables. Among the guests was J. Liddon Pennock Jr., who was already in the habit of purchasing four sheep each year from Widener's Erdenheim.[79]

Beyond this Widener was known for his generosity throughout both the racing community and the Philadelphia area. Giving away scads of money to worthy organizations as a multimillionaire is one thing—and the Wideners did it in spades—but giving to people, to individuals is quite another. In this regard, Widener set a fine example to locals during World War II by opening Erdenheim every weekend to wounded soldiers quartered at the Valley Forge General Hospital; these troops were treated as the Wideners' personal guests. It is estimated that more than twenty-eight hundred patients were entertained in this fashion during the course of the war.[80]

The Old Quarries, The Elkins Parcel

At various times during his ownership of Erdenheim, Widener parted with parcels he found no longer fit with the mission of Erdenheim Farm. In 1938, for example, Widener sold two small lots at the north end of the old Daniel Williams tract to G. and W. H. Corson Incorporated, presumably to be used as additional quarry storage space. Eleven years later, on March 24, 1949, the Wideners sold another lot, this time of 16.347 acres, to the same company for a nominal fee; these transactions effectively trimmed the northern end of the Williams lot (or present-day Sheep Tract) considerably and resulted in the old Williams quarrying operations being entirely in the hands of G. and W. H. Corson Incorporated.[81] For similar reasons Widener disposed of an old quarrying operation at the south end of the old Erdenheim parcel, along Wissahickon Avenue in 1954. For the nominal fee of $1.00, the Wideners transferred 3.07 acres to the Township of Springfield; today, this land remains in control of the Township of Springfield, which uses it as a highway maintenance facility and a recycling location.[82]

Perhaps the most significant of these land sales, however, was Widener's decision to part with a substantial portion of the old Kittson tract that lay on the west side of Thomas Road (then called Fullers Road). On January 9, 1931, the Wideners transferred 41.637 acres to Elizabeth Tuckerman Elkins—land that was located directly opposite the Thomas Road entrance to "New" Erdenheim.[83] Following this sale, Widener's cousin, William McIntire Elkins, built the stately mansion Briar Hill for his wife, Elizabeth, who had purchased the property. Georgian Revival in design, Briar Hill included a central residence with two projecting wings on either side of a massive courtyard. Following William McIntire Elkins's death in 1947, Briar Hill was sold to Rosabelle Deichelmann; she and her husband, Dr. S. J. Deichelmann, soon transformed Briar Hill into Eugenia Hospital, a mental health facility that finally closed in 2004, at which time Briar Hill

FACING: Elkins parcel with Briar Hill, on the west side of Thomas Road opposite the entrance to "New" Erdenheim, 1938. Detail from *Property Atlas of Montgomery County, Pa.,* volume D, Philadelphia: Franklin Survey Company, 1938, plate 20, Map Collection, Free Library of Philadelphia.

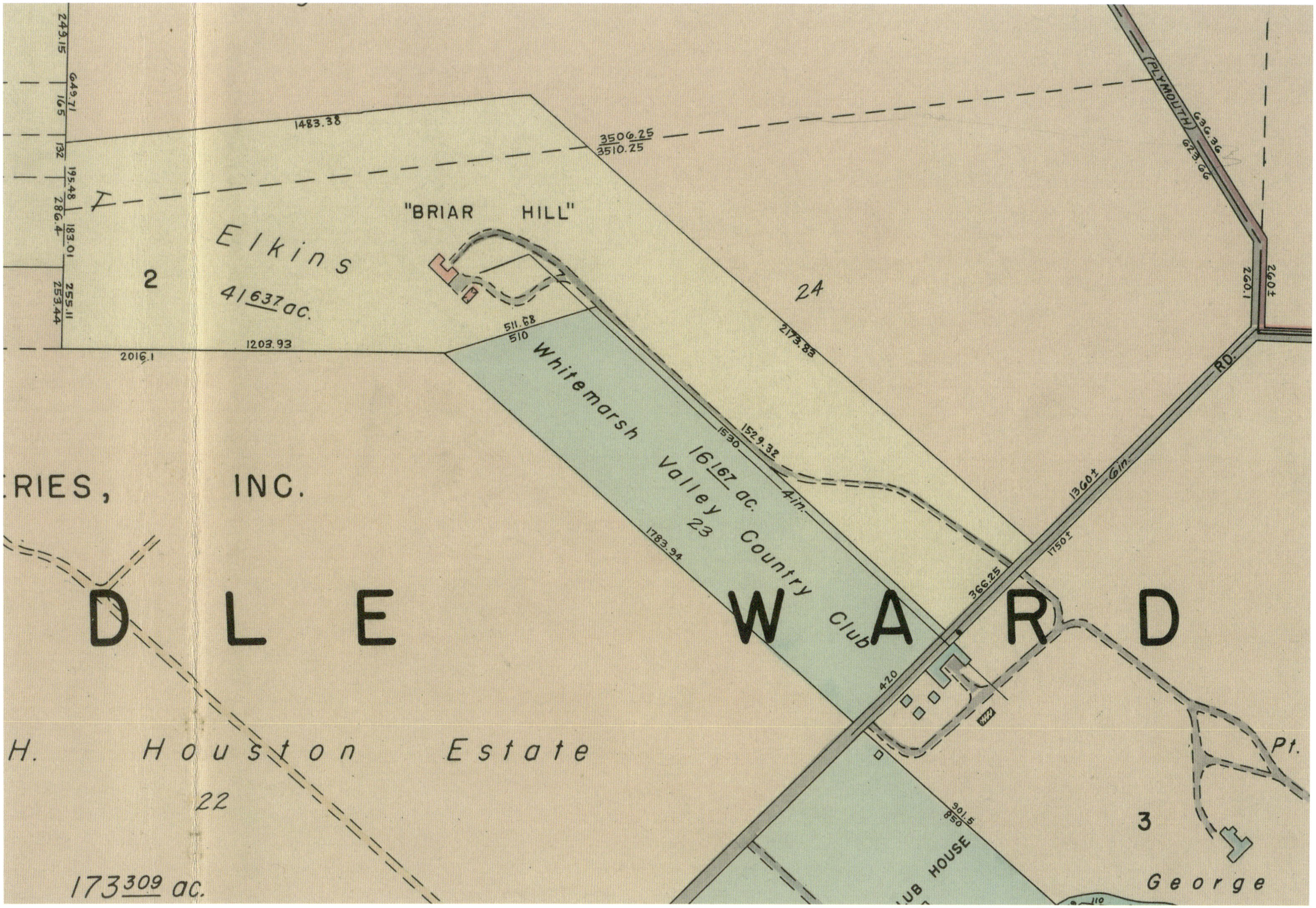

"BRIAR HILL"
Elkins
2
416.37 ac.
2A
Whitemarsh Valley Country Club
161.67 ac.
23
RIES, INC.
DLE WARD
H. Houston Estate
22
173.309 ac.
(PLYMOUTH)
RD.
CLUB HOUSE
3
George
Pt.
1483.38
3506.25
3510.25
2173.83
1203.93
2016.1
511.68
510
1529.32
1530
1783.94
366.25
1750±
1360±
420
901.5
850
636.36
623.66

was demolished. Today the grounds are home to a retirement community, The Hill at Whitemarsh. Portions of the Briar Hill mansion were auctioned off prior to its demolition, however, and Elizabeth Elkins gave the entire library and its contents, including its furnishings, to the Free Library of Philadelphia. The room is now a part of the Rare Book Department at the Central Branch.[84]

Widener's Death, and the Future of Erdenheim

When George Widener Jr. died on December 8, 1971, it was a stunning blow to both the racing community and the intellectual community, despite Widener's long illness and advanced age. For many years a philanthropist whose support went to the arts, the sciences, and education, Widener had served at various points in his life as a trustee for the Zoological Gardens and Philadelphia's Academy of Natural Sciences, as well as a director of the Smithsonian's National Gallery of Art and chairman of the Philadelphia Museum of Art up to the time of his death. He was an ardent supporter of several major visual arts and performing arts institutions at Saratoga Springs, Philadelphia, and New York, and his support helped endow the Widener Library (named for his brother who died on the *Titanic*) at Harvard University.[85] His presence was still equally palpable in the racing world. At the time of his death, he remained chairman of the National Museum of Racing in Saratoga Springs, New York, and the honorary chairman of both the New York Racing Association and the Jockey Club. He was renowned for the numerous innovations he had made in the sport, including the introduction of a training school for race officials, a revision and simplification of the racing silk registration process, new classifications for blood typing of thoroughbreds, and new methodologies for horse breed identification. Nevertheless Widener's ill health had been a cause for concern for quite some time. In 1970, for example, a year before his death, the National Museum of Racing awarded Widener its first "Exemplar of Racing" prize—a magnificent and well-deserved honor—but Widener was not capable of attending.[86]

Widener's death led to the divestiture of much of the stock at Erdenheim and Spendthrift. Most notable among these sales was the world record price paid for What a Treat, one of Widener's most prized brood mares and alleged by some to be his favorite of all the horses he raised; a French operation put down $450,000 for just that single horse.[87] All told, the estate planned four dispersal sales, the first occurring on February 14, 1972, in Kentucky, when twenty-nine brood mares and fifteen yearlings were sold at auction.[88] This one dispersal event alone was indicative of Widener's success as a breeder. Eleven descendants of one filly—Rare Perfume—brought the astounding sum of $1,545,000 (nearly a third of which was the sale of What a Treat). Among the purchasers was F. Eugene Dixon Jr., Widener's nephew, who purchased two yearlings that he intended to keep at Erdenheim following the settlement of the estate. As one report described it, "In style, setting, bidders . . . , and quality of horses offered, the Widener dispersal defies comparison." To contextualize its magnitude further, twenty-five mares brought an average of $99,120 apiece, far surpassing the previous record from the 1966 DuPont dispersal of $60,954 apiece for eighteen mares.[89]

In early March 1972, the estate followed up the astounding success of the first dispersal auction with the sale of fifteen of Widener's two year olds, who averaged $125,833 apiece, with Treat Me Fair (purchased by Dixon) bringing $410,000—a record for a two year old in training.[90] This sale was followed in mid-March by another, where Bendara, the three-year-old daughter of Never Bend, matched the record set just a week earlier by Treat Me Fair. The total of all three sales brought over $6.6 million for the estate, at an average of $96,286 for the sixty-nine animals.[91] Such high-dollar sales may indeed have been partially a result of what was termed a "free speech policy" that Widener had insisted be adopted at the sale of his estate. All questions about the conditions and health of the horses were answered in full, regardless of whether such facts might diminish the price that bidders would otherwise be willing to pay. Some argued that Widener's example would never eliminate the practice of caveat emptor that usually ruled the day at such events; others simply observed that the practice was reflective of the man and his ethics that they had known in racing circles for more than fifty years.[92] Beyond these sales the Widener estate also offered options on the future seasons of its remaining stallion stock; by the end of February 1972, eleven of the twenty options had been sold.[93]

Widener's death also left the question of what to do with his vast fortune. Under the terms of his trust, Widener instructed that much of his money be given "to benefit charitable, religious, educational, literary, and scientific purposes and uses," thus continuing exactly the kinds of philanthropic pursuits he had engaged in during his lifetime. The principal figure responsible for determining the distribution of these trust funds was Fitz Eugene Dixon Jr., Widener's nephew and primary heir, given that the Wideners had no children.[94] Once the estate was settled, it also fell to Dixon to figure out what to do with Erdenheim, given that much of the stock was now gone. Over the thirty-five years that followed, Dixon would prove that he was up to this task.

Notes

1. Thomas H. Bateman, compiler, *Widener-Dunton-Elkins-Broomall and Allied Families* (New York: American Historical Company [privately printed], 1953), 1–4; Frank Fitzpatrick, "Philly's Horse Racing Heritage," *Philadelphia Inquirer Online,* May 19, 2011, http://www.philly.com/philly/blogs/frank_fitzpatrick.
2. Frederick Platt, "Horace Trumbauer's Mansions in Elkins Park," *Bulletin of the Historical Society of Montgomery County,* 30:4 (Spring 1997), 272, and Bateman, *Widener,* 4–5.
3. Platt, "Horace Trumbauer's," 273–77.
4. Bateman, *Widener,* 5–6; J. Thomas Jable, "Widener, George Dunton, Jr.," *American National Biography,* edited by John A. Garraty and Mark C. Carnes, volume 23 (New York: Oxford University Press, 1999), 336–37.
5. Fitzpatrick, "Philly's," and Platt, "Horace Trumbauer's," 272.
6. Platt, "Horace Trumbauer's," 281.
7. O'Neil Sevier, "Breeding Growing Stronger in the East," *Philadelphia Inquirer,* March 4, 1917, 22; William E. Fischer Jr., "Widener, George Dunton," *Dictionary of American Biography,* Supplement 9: 1971–1975, edited by Kenneth T. Jackson (New York: Charles Scribner's Sons, 1994), 866.
8. Quoted in Jable, "Widener," 337.
9. Buckley purchased this tract in 1915 from the Real Estate Title Insurance and Trust Company of Philadelphia, which handled the settlement of the Carson estate. It is likely that Buckley paid only a nominal sum. For the Widener transaction, see R. Nelson Buckley to George D. Widener, May 15, 1916, Deed Book 745, page 79, Montgomery County Archives. Two later transactions were necessary to clean up the boundary with Carson College, involving minimal transfers of land. For these, see R. Nelson Buckley, etc., to George D. Widener, June 18, 1926, Deed Book 997, page 92, and George D. Widener, et ux., to R. Nelson Buckley, etc., June 18, 1926, Deed Book 997, page 104, Montgomery County Archives.
10. Real Estate Title Insurance and Trust Co., et al., Trustees, to George D. Widener, May 15, 1916, Deed Book 745, page 83, Montgomery County Archives. This transaction with Widener makes no mention of the old burial ground plot, which lay within these boundaries. On the Whitemarsh Valley Country Club outparcel, see *Property Atlas of Montgomery County, Pa.,* volume D (Philadelphia: Franklin Survey Company, 1938), plate 20.
11. Alfred K. Williams, et al., to John Faber Miller, September 13, 1916, Deed Book 748, page 414, Montgomery County Archives, and John Faber Miller, et ux., to George D. Widener, September 18, 1916, Deed Book 746, page 482, Montgomery County Archives. On the two smaller lots excluded, see Daniel Williams to Plymouth Railroad Company, February 9, 1869, Deed Book 170, page 272, and Daniel Williams to Pennsylvania Railroad Company, March 21, 1892, Deed Book 367, page 6, Montgomery County Archives.
12. Clifton S. Hunsicker, *Montgomery County, Pennsylvania: A History,* volume 3 (New York: Lewis Historical Publishing, 1923), 646–47.
13. Roberts, "Walter Coulston," *Biographical Annals,* 74–76.
14. Walter Coulston, et al., to John Donahue, December 30, 1916, Deed Book 753, page 142, Montgomery County Archives. Deeds conveying this tract and the Corson tract to Widener could not be located.
15. Albert S. Schmitz, et ux., to George D. Widener, February 23, 1926, Deed Book 977, page 403, Montgomery County Archives. The second parcel in this transaction was not part of Erdenheim.
16. "The Erdenheim Stud," *Field Illustrated,* March 1925, quoted in notes taken by William H. Weber Jr., Springfield Township Historical Society, Folder 2005.4.141. It is well documented

that Dr. John J. McCloskey was at the stud under Carson (taking his father's place), but A. J. McCloskey was living in the vicinity in 1920.

17. "Erdenheim 'Chase to Fenwick Entry," *Philadelphia Inquirer,* October 10, 1920. In After a brief, dismissive announcement of the winners, the vast majority of this article focused on who was present and what they were wearing.
18. "Erdenheim Races Draw Big Crowd, *Philadelphia Inquirer,* October 16, 1921.
19. The White Marsh Valley Hunt Club to George D. Widener and Schofield Andrews, March 25, 1954, Deed Book 2456, page 38. R. Nelson Buckley sold this parcel, which had been leased by Carson for a ten-year period in 1907, to the hunt club in 1917. See R. Nelson Buckley to White Marsh Valley Hunt Club, April 2, 1917, Deed Book 756, page 220, Montgomery County Archives.
20. The Williams buildings and the stone crusher were gone by 1927, while the houses at the corner of Stenton and Wissahickon were demolished between 1927 and 1938.
21. This is the house in which Bill Weber and his father, William H. Weber—one of Erdenheim's gardeners after 1926 and its superintendent of grounds and greenhouses from 1949 to at least 1961—is believed to have lived for a time. See Bill [Weber] to "Marie," undated note in Folder 2005.4.141, Springfield Township Historical Society, discussed in more detail below. For more on the elder Weber and Erdenheim, see William H. Weber, "Year-Round Flower Production at Erdenheim Farm," *Garden Journal* 11–12 (1961): 16–17.
22. See Bill [Weber] to "Marie," undated note in Folder 2005.4.141, Springfield Township Historical Society. In his note Weber explains that the 1764 stone is visible in the chimney at the time he wrote his note (probably in the 1980s), but a 1902 image of the house (not found with Weber's materials, but perhaps the image by Moses King) indicated that this chimney was not on the house at that time. Weber concludes, "Widener's renovation probably put it there in a new chimney."
23. "The Erdenheim Stud," quoted in Weber, "Year Round," 16-17. The correspondence of Olmsted Associates with Widener regarding changes at Erdenheim during the 1920s frequently mentions Widener's deferral to Mrs. Widener's wishes and his efforts to minimize disruptions on the property during periods when she was staying at the home. See Olmsted Associates Records, Job Files, 1863–1971, box B417, reel 373, project 7378, Library of Congress.
24. "Horse Racing Giant Widener, Dies at 82," *Lexington Herald,* December 9, 1971.
25. Widener purchased five separate South Carolina plantations in the Pocotaligo area sometime during the 1920s and by 1930 had combined them into the single plantation known as Mackay Point. Prior to these purchases, Widener hunted frequently and extensively in the area. At one point, Widener agreed to publish his quail and duck recipes. See clipping from Mary Frost Mabon, "Mr. Widener's Quail," *Sports Illustrated,* October 6, 1958.
26. E.M., "Local History Sketch," *Ambler Gazette,* September 1, 1904, 2.
27. Yeakle, "Old Burial-Ground," 102–3. A review of the deed relevant to this transaction shows no reference to the burial ground in question. See "Real Estate Title Insurance and Trust Co., et al., Trustees, to George D. Widener," Montgomery County Deeds, Book 745, page 83, Montgomery County Archives. The cemetery was briefly described as still being extant in 1907, "surrounded by a clump of pines." See "Where Robert N. Carson's Millions Will Rear a College for Orphan Girls," *Public Ledger,* October 24, 1907.
28. "Abandoned School Sold," *Philadelphia Inquirer,* November 28, 1921, 2.
29. H. S. Finney, "Erdenheim," *Blood-Horse,* June 22, 1935, 746.
30. These two new buildings first become evident on *Atlas of the Reading Main Line, Montgomery County, Pennsylvania,* volume 1 (Philadelphia: Frank H. M. Klinge, 1927), plate 23. Finney, "Erdenheim," 747.
31. George D. Widener, et ux., to Commonwealth of Pennsylvania, November 15, 1920, Deed Book 822, page 575, Montgomery County Archives.
32. Olmsted Brothers, "Revised Estimate of Cost of Revising Gardens," February 24, 1925, Olmsted Associates Records, Job Files, 1863–1971, box B417, reel 373, project 7378, Library of Congress. All documents (not images) from this project are at this location, while all images are held at the Olmsted Archives, Frederick Law Olmsted National Historic Site, Brookline, Massachusetts. For simplicity's sake, full citations referencing repository location will not appear hereafter.
33. Horace Trumbauer to Olmsted Brothers, April 3, 1925, and Jacob Sloet, "Some of the Things Discussed on the Ground with Mr. Gallagher and Mr. Graham," July 22, 1925, Olmsted Associates Records. Percival Gallagher of Olmsted Brothers appears to have been Widener's principal contact on the projects.
34. Olmsted Brothers to Harold Graham, May 25, 1925, Olmsted Associates Records.
35. Sloet, "Some of the Things," Olmsted Associates Records.
36. Sloet, "Some of the Things," Olmsted Associates Records.
37. Horace Trumbauer to Olmsted Brothers, February 23, 1926.
38. Olmsted Brothers to George D. Widener, August 11, 1925, Olmsted Associates Records. Widener's on-site gardener at the time appears to have been named "Starkie," perhaps a nickname for Starke, as seen in other correspondence on the project. See Olmsted Brothers to George D. Widener, December 14, 1925, Olmsted Associates Records.
39. Finney, "Erdenheim," 747.
40. Gallagher to George D. Widener, August 14, 1931, Olmsted Associates Records.
41. "5000 See Firemen Fight $40,000 Blaze on Widener Estate," *Philadelphia Inquirer,* January 17, 1921.
42. "5000 See Firemen."
43. Fitzpatrick, "Philly's Horse"; Jable, "Widener," 336; "George D. Widener, 1889–1971," *Blood-Horse,* December 20, 1971, 4764.
44. "Can He Really See That Far Ahead?" uncredited, undated [1962] clipping, Widener clippings file, Keeneland Library.
45. Jable, "Widener," 336.
46. Sevier, "Breeding."
47. "George Widener, Giant of Thoroughbred Racing, Succumbs," *Mobile Register,* December 9, 1971, 76; Fischer, "Widener," 866.

48. John Hervey, *Racing in America, 1922–1936* (New York: Scribner Press, printed privately for the Jockey Club, 1937), 35–37.
49. "Jamestown to Vie in Stakes at Arlington," *Times-Picayune,* March 7, 1931, 12; "Jamestown," *Turf and Sport Digest,* October 1930.
50. Hervey, *Racing in America, 1922–1936,* 120–21.
51. Hervey, *Racing in America, 1922–1936,* 160–61.
52. Hervey, *Racing in America, 1922–1936,* 257.
53. "George D. Widener," undated obituary, *Chronicle of the Horse,* 8, Widener clippings file, Keeneland Library.
54. "Wide Open Race Is Anticipated in 70th Kentucky Derby on May 6," *Trenton Evening Times,* February 16, 1944, 12.
55. Albion Hughes, "Forty Years of Improving the Breed," *Sunday Bulletin Magazine,* April 19, 1964, 6–7.
56. "George D. Widener, 1889–1971," *Blood-Horse,* December 20, 1971, 4764.
57. Robert F. Kelley, *Racing in America, 1937–1959* (New York: North River Press, printed privately for the Jockey Club, 1960), 151–52.
58. "Turf Solon George D. Widener Dies Following Long Illness," *Daily Racing Form,* December 9, 1971.
59. "George D. Widener," *Blood-Horse,* December 20, 1971, 4764.
60. "Mrs. Widener Dies in Florida," No Publication, March 12, 1968, Widener clippings file, Keeneland Library; "Horse Racing Giant Widener, Dies at 82," *Lexington Herald,* December 9, 1971.
61. "Widener, George Dunton," *National Cyclopedia of American Biography,* volume 57, Clifton, N.J.: James T. White, 1977, 264.
62. "George Widener, Giant of Thoroughbred Racing, Succumbs," *Mobile Register,* December 9, 1971, 76.
63. "Widener," *National Cyclopedia,* 264.
64. "Is Landis Wrong in His Stand Against Racing in Baseball?" *Augusta Chronicle,* September 10, 1943, 15.
65. "Widener," *National Cyclopedia,* 264.
66. Jable, "Widener," 337.
67. "Widener," *National Cyclopedia,* 264.
68. Jable, "Widener," 337.
69. Neil Newman, "Jack Joyner's Life Story," *Blood-Horse,* December 18, 1943, 796–97; "A. J. Joyner Dies at Erdenheim," *Blood-Horse,* September 11, 1943, 311–12.
70. "Album of the Turf," *Blood-Horse,* June 19, 1943, 772.
71. "General Items," *Blood-Horse,* August 28, 1943, 271; "A. J. Joyner Dies," 312.
72. Neil Newman, "Jack Joyner's Life Story," *Blood-Horse,* May 27, 1944, 751.
73. For a lengthy treatment of Andrew Jackson Joyner's life, see Neil Newman, "Jack Joyner's Life Story," *Blood-Horse,* December 18, 1943-February 3, 1945.
74. Joe H. Palmer, "Mulholland and His Horses," *Blood-Horse,* November 20, 1943, 620.
75. "Trainer Bert Mulholland Suffers Coronary Occlusion," *Daily Racing Form,* August 4, 1967; "Famed Trainer Passes," *Lexington Herald Leader,* July 13, 1968; "Elect Trainer Bert Mulholland to Hall of Fame at Saratoga," *Daily Racing Form,* May 27, 1967; "Bert Mulholland on Turf Scene since 1899," *Daily Racing Form,* July 24, 1965; Albion Hughes, "Forty Years of Improving the Breed," *Sunday Bulletin Magazine,* April 19, 1964, 6–7.
76. Kent Hollingsworth, "Fillies and Blue Silks," *Blood-Horse,* March 20, 1972, 969.
77. "Erdenheim Farm" advertisement, *Field Illustrated,* December 1920, 1167.
78. "Jerseys Average $1000 at Sale," *Philadelphia Inquirer,* June 6, 1919.
79. Ruth Seltzer, "Widener Entertains at Showplace Farm," *Philadelphia Inquirer,* October 27, 1968.
80. "Horse Racing Giant Widener, Dies at 82," *Lexington Herald,* December 9, 1971.
81. George D. Widener, et ux., to G. & W. H. Corson, Inc., March 1, 1938, Deed Book 1269, page 217, and George D. Widener and Jessie Widener to G. & W. H. Corson, Inc., March 24, 1949, Deed Book 1989, page 466, Montgomery County Archives.
82. George D. Widener and Jessie Widener to Township of Springfield, July 2, 1954, Deed Book 2484, page 191, Montgomery County Archives.
83. George D. Widener, et ux., to Elizabeth Tuckerman Elkins, January 9, 1931, Deed Book 1122, page 94, Montgomery County Archives.
84. Rachel Hildebrandt, *The Philadelphia Area Architecture of Horace Trumbauer* (Mount Pleasant, S.C.: Arcadia Publishing, 2009), 84–85; Larry Fish, "Designed by Trumbauer, Destined for Destruction," *Philadelphia Inquirer,* November 19, 2004.
85. Fischer, "Widener," 867.
86. Jable, "Widener," 337; "George Widener, Giant of Thoroughbred Racing, Succumbs," *Mobile Register,* December 9, 1971, 76.
87. "Widener," *National Cyclopedia,* 264; Joe Hirsch, "Tropical Park," *Daily Racing Gazette,* December 23, 1971.
88. "Special Breeding Stock Dispersal for Widener Estate Scheduled Here," *Lexington Herald,* January 7, 1972.
89. "What's Going on Here," *Blood-Horse,* February 21, 1972, 661; Rab H. Hagin, "More Francs Than Yen—and Plenty of Dollars," *Blood-Horse,* February 21, 1972, 662, 664.
90. "$410,000 for a 2-Year-Old Filly," *Blood-Horse,* March 13, 1972, 882.
91. Kent Hollingsworth, "Fillies and Blue Silks," *Blood-Horse,* March 20, 1972, 968.
92. Hugh J. McGuire, "Monday's Widener Dispersal Will Reflect Policy of Man," *Thoroughbred Racing Form,* February 14, 1972.
93. "Widener Stallion Seasons Selected," *Blood-Horse,* February 28, 1972, 746.
94. An account of the distribution of these trust funds can be found in David Lewis Eynon, *A Trust in the Blood,* Philadelphia: Provident National Bank, 1987.

CHAPTER SIX

Fitz Dixon and the Legacy of Erdenheim

Born at his family's summer estate on Grindstone Neck in Winter Harbor, Maine, on August 14, 1923, Fitz Eugene Dixon Jr. took several unexpected paths with his life and pursued interests that seem to have changed course on frequent occasion. Dixon spent his youth at the family's home, Ronaele Manor, in Elkins Park, another of the majestic mansions built by Horace Trumbauer for the Widener-Elkins-Dixon families, but all indications are that he disregarded his uncle's persistent advice to become a doctor or a lawyer, something remunerative and befitting his place in society. Despite his parents' divorce, young Fitz remained with his mother at Ronaele Manor through his secondary education, and whether it was the experience of being the child of divorce or some other influence that made him think differently about his wealth is not clear. Shortly after completing this education at Episcopal Academy, though, Dixon reluctantly enrolled at Harvard, but after seven months, he returned to Episcopal Academy to work as a teacher, in spite of the fact that he did not possess a college degree. Over the next two decades, he taught French, English, history, and health, as well as coaching the squash team and working for a time as the director of athletics and an assistant headmaster. It was a period he once described as "the happiest days of my life," in large part because he actually worked for a living, something he claimed at various times was his unique place among the Wideners and Dixons of the twentieth century. In his personal life, he followed a more pedestrian path, too. In his late twenties, in 1952, Dixon married Edith Bruen Robb, the sister of one of his students at Episcopal, and the couple had two children, George Widener Dixon and Ellin Dixon Miller.[1]

By some accounts, Dixon was "impatient at times, irritable . . . , compulsive, quick to blow up. But generous."[2] Whether or not this stemmed from his unique position as the sole heir to not just one but two massive family fortunes (Elkins and Widener) is not clear. Never once, though, did Dixon stand on his elevated position and buy into the illusion that teaching at his alma mater was a position that was beneath him. Asked about it in 1976, he acknowledged, "I'm fortunate enough that I don't have to go out there and earn a paycheck, but I couldn't sit home and do nothing. . . . I'd be a martini drunkard at the end of six months."[3] On other occasions, having so much money—whether in association with the dispersal of the Widener estate or as a result of his own investments—was something Dixon regarded as a bit of a curse, in large part because he

FACING: Stone Bridge over the Wissahickon, Erdenheim. Photo by Eric Plaag, May 2012.

believed people loved him for his money rather than who he was as a man. "Sometimes," Dixon told one reporter in 1978, "I wish I didn't have all this bleeping money. . . . I just hope that when they put me in the sod, a few people will shed a few tears and somebody will say, 'Yes, sir, that Fitz Dixon, he was a pretty good boy.'"[4] When his daughter Ellin elected not to debut, as was the social custom of the day, Dixon was reportedly ecstatic, saying, "The days of big social bashes are gone. It's foolish to fritter away money like that. Foolish." And yet, as a constant reminder of his connection to his ancestors and the origin of his wealth, Dixon frequently wore an emerald ring given by his grandfather George D. Widener Sr. to his grandmother as she stepped into a lifeboat on the *Titanic*.[5]

In addition to his other interests, Dixon, like Widener, was a man of the turf. At the dispersal of the Widener estate, Dixon bid heavily (some would say wildly) under the public perception that he was driving up bids for money that would ultimately come back to him as the sole heir of the estate. Others noted that the Widener dispersals were designed to generate money for the philanthropic plans of the estate, and that Dixon's expenditures were from his own money, not from what he would be inheriting from Widener. Beyond his purchases of horses, however, Dixon was also a member of the Jockey Club, served as president of the Horsemen's Benevolent and Protective Association, and was for a time the director of Hialeah. While his horse stock holdings were substantially smaller than Widener's, Dixon nevertheless saw champions from his stable in Will I Rule, Wageko, and Air Delight. Following Widener's death, he raced almost exclusively under the Erdenheim banner, although he also competed horses in dressage and steeplechase.[6] His real fame on the turf, however,

Dixon Family Crest, from Thomas H. Bateman, compiler, *Widener-Dunton-Elkins-Broomall and Allied Families,* 1953.

probably came with his jumpers, most notably Jet Run, with whom he won both team and individual gold medals at the 1979 Pan American games and the 1981 FEI Jumping World Cup Final. Jet Run was also named AGA Horse of the Year in 1982. Dixon's contributions to the Devon Horse Show in the 1960s led to the erection of a new show ring, now known as the Dixon Oval.[7] Dixon also served for many years on the Pennsylvania Horse Racing Commission, including as its chairman.[8] Dixon was also part of the syndicate that purchased Secretariat after his Triple Crown victory and sent him to stud.[9]

Dixon's most enduring, national fame, however, probably lies in his ownership of the Philadelphia 76ers professional basketball team in the late 1970s, during which time he signed the immortal Julius "Dr. J" Erving.[10] After paying $8 million for the team in 1976, Dixon promptly went out and lured Erving to Philadelphia for another $6.6 million. During his tenure the Sixers advanced to the NBA championship twice, but by 1980 Dixon's run with the team had soured, and he sold his interest.[11] Those years were marked by controversy, and they left Dixon ambivalent about his foray into team sports ownership. He was frequently vilified in the press for his alleged incompetence as the owner of a basketball team; after he fired the coach who led the Sixers to that first championship round six games into the next season, the local press wrote, "So, once again, we learn that you can go to Harvard, build a few ships, wear polished loafers and still confuse manure with tuna fish." Dixon responded by noting that he had never built any ships.[12]

Some, including Dr. J, found his money intimidating. Others alleged, "He's not an evil man. He just doesn't know how to behave." Indeed his other efforts in

Fitz Eugene Dixon Jr., Templana Collection, Special Collections Research Center, Temple University Libraries, Philadelphia, Pa.

the professional team sports world seemed to hint at his being uncomfortable in those arenas, despite repeated attempts during his life to make his mark. He also possessed ownership stakes at various times in the Philadelphia Phillies (1940s), Philadelphia's professional hockey team during the 1950s, and perhaps most notably the Philadelphia Eagles, whom he owned shares of briefly, then bid on but refused to purchase outright when the price seemed too high.[13]

Dixon as Erdenheim's Caretaker and Philanthropist to the World

In the years following Widener's death, Dixon shepherded the initial efforts to preserve what remained of historic Erdenheim Farm. As part of the settlement of Widener's estate, for example, a 117-acre portion of the farm on both sides of the Wissahickon and adjoining Stenton Avenue was given to the Natural Lands Trust, including the original house location of "Old" Erdenheim. The remainder of the farm holdings, which were left to Dixon, continued to function as a "gentleman's farm," where Dixon continued the age-old pursuits of raising and showing Angus cattle, thoroughbred horses, and Cheviot sheep.[14] One example of this success was Erdenheim's performance with its Cheviot sheep at the 1972 Eastern States Exposition, where Erdenheim took home six first-place ribbons for yearling rams, yearling ewes, best pair of ewes, junior ewe lambs, exhibitor's flock, and breeder's young flock.[15] That same year the Erdenheim flock of Cheviots took home twelve of the thirteen first prize ribbons at the Great Barrington Fair in Massachusetts.[16] Dominant performances like that under Dixon's ownership of Erdenheim were not unusual.

Like his predecessors, the Wideners, Dixon also raised flowers at Erdenheim and was so successful with these efforts that *Philadelphia* magazine speculated in 1974 that the floral plantings on the farm alone were worth more than $1 million.[17] Among locals, however, Dixon was renowned for the care he gave to the property at Erdenheim, endeavoring to keep its appearance spotless and to repair the inevitable damage from storms or decay immediately. "One day you noticed the damage while driving through the farm, the next day it had either disappeared or been repaired," one local resident observed. "It sometimes was easy to envision a tribe of munchkins roaming the property at night, making all things right before the sun rose the next morning."[18]

Away from Erdenheim and the sports world, Dixon was also a giant in education and various philanthropic pursuits. Early in his career, when barely in his thirties, Dixon played a crucial part in the establishment of the Maine Coast Memorial Hospital in Ellsworth, Maine, often covering the payroll with his personal funds in order to ensure that the hospital would remain open.[19] Beginning in 1960 Dixon joined the board of trustees for Pennsylvania Military College, which became Widener College in 1972, then served as its board chairman from 1972 to 1984 and again from 1987 to 1998.[20] In addition to his work on behalf of Widener College (now Widener University), he served on Temple University's board of trustees from 1970 to 1983 and as its chairman during the last six years of his tenure. During this period he was especially focused on Temple's Ambler campus, where he facilitated construction of two classroom buildings subsequently named Widener Hall and Dixon Hall. His 1969 donation of a thirty-room manor in Chestnut Hill—named the Eleanor Widener Dixon House for his mother—allowed Temple to expand its Albert M. Greenfield Conference Center operations. He was the first chairman of the board of governors for the State System of Higher Education in Pennsylvania, and he also served as the chairman of the Delaware River Port Authority. Dixon also served as president and trustee of the Widener Memorial Foundation in Aid of Handicapped Children in Lafayette

Hill, and as a board member for the Abington Memorial Hospital, the Fairmount Park Commission, and the Germantown Hospital and Medical Center.[21]

On the arts front, Dixon was a board member for the Philadelphia Museum of Art and played a critical role in returning Philadelphia's iconic "LOVE" sculpture to the city when its artist, Robert Indiana, loaned the sculpture to the city for the Bicentennial, then reclaimed it after the city refused to pay the asking price of $45,000. Dixon offered the sculptor $35,000, which Indiana accepted, then gave the sculpture back to Philadelphia.[22] Back in Maine, where he frequently vacationed throughout his life, Dixon funded Schoodic Point's renovation of base housing that was abandoned by the U.S. Navy in 2002, and he purchased a popular restaurant and inn in Hancock, Maine—Le Domaine—in 2005 to ensure that its doors would remain open.[23]

Development Comes to Erdenheim

For decades after Widener's death, local residents wondered what would become of the rest of Erdenheim, beyond the portion set aside by Widener for the Natural Lands Trust. Some hoped it would always remain the historic farm they had come to love and drove past as a diversion on their way to work. Others saw dollar signs in its sod, a place for McMansions and mini-malls. In 2001, when Dixon arranged to sell 54 acres of Erdenheim land to the developers who had taken over the old Elkins parcel on which Briar Hill and, subsequently, the Eugenia Hospital, had sat, the biggest hopes of speculators and the worst fears in the hearts of Erdenheim's neighbors seemed realized. Development plans called for the creation of 150 cottages, 175 apartments, and a large assisted-living and health care facility on the Erdenheim and Eugenia parcels, the entirety of which would be called The Hill at Whitemarsh, with much of it located along Thomas and Flourtown Roads, covering portions of the Angus Tract. At zoning meetings held early that year, more than four hundred residents turned out in opposition to the proposed plan, backed by the Whitemarsh Valley Farms Community Association, which was adamantly opposed to the proposal.[24] Prized as the linchpin of the remaining open green space stretching from Fort Washington to Chestnut Hill, Erdenheim's pristine, pastoral fields became a preservation priority for many local residents. When the Whitemarsh Township Planning Commission subsequently rejected the proposal in March 2001, many in the local community were relieved.[25]

The developers, fronted by John J. Fleming, did not back down. Instead they found a way to work with Dixon and local residents to find a compromise. As part of the deal, Fleming agreed to rework the development plans and tuck the entire complex just behind a ridge running parallel to upper Flourtown Road in order cut down on the site's visibility from the farm. In addition he agreed to help create and fund the Whitemarsh Foundation, an organization committed to preserving Erdenheim Farm's remaining land. While some neighbors expressed misgivings about the funding terms of the deal (which would have permitted the agreement to be vacated if the other parcels were not purchased for preservation in a timely fashion, and would have allowed Fleming to purchase both the Angus and the Sheep Tracts for large-scale development), others were buoyed by Fleming's willingness to work with the community and his promise to preserve 12 acres of woodland as a buffer between the farm and the retirement community. Estimates suggested that at least $20 million would be needed to purchase the remainder of Erdenheim not protected by the Natural Lands Trust, and to kick start that fundraising effort, Fleming agreed to donate $500,000 up front and up to an additional $600,000 under certain circumstances.[26]

Part of Dixon's push to see the Eugenia parcel and a portion of Erdenheim go to developers was that he had grown weary of inmates from the Eugenia psychiatric facility wandering onto Erdenheim's land.[27] Using this concern to define the Eugenia tract as a "blighted area," the Montgomery County Redevelopment Authority came up with a novel "tax increment financing" scheme to funnel up to $1.1 million in new property taxes on the proposed development into preservation of the remaining Erdenheim land. Nevertheless local residents remained skeptical of the developer's intentions.[28] Over the next few months of 2003, however, other partners signed onto the deal, including the Colonial School District, which committed up to $11.6 million in future tax revenues (coming primarily from the operations of The Hill at Whitemarsh) for preservation of Erdenheim. In exchange the Erdenheim property would lose its special tax consideration as farmland, generating an additional $150,000 per year that would go to the schools, plus another $85,000 that would go to the township and the county.[29]

In December 2004 development moved forward at the old Eugenia property, including a groundbreaking ceremony in which both Fitz Dixon and his wife, Edith, participated.[30] At about this same time, Montgomery County officials began pushing for the creation of the Wissahickon Green Ribbon Trail, an effort to complete the continuous Fairmount Park trail from Fort Washington to the heart of Philadelphia that had originally been proposed in the 1890s. The missing link in Montgomery County's plans was the .75-mile section of Erdenheim Farm through which the Wissahickon flowed. The easement the county sought from Dixon would have created an eight- to ten-foot-wide multiuse path along the creek, but Dixon was still reluctant to sign on as late as April 2005.[31]

FACING: Cattle watering at the Angus Tract, just north of the Coulston House. The Hill at Whitemarsh complex is visible at upper left. Photo by Eric Plaag, May 2012.

Following Dixon's death from melanoma in August 2006, pressure to preserve the farm mounted as developers still licked their chops at the prospect of obtaining the land, given that the clock was now ticking on the one-year option period the Whitemarsh Foundation had to purchase the Angus Tract. Recognizing Erdenheim as one of the few remaining open areas of historic farmland in Montgomery County, and perhaps the last in the Wissahickon Valley, the county commissioners realized that their help was needed, too. Toward this end Montgomery County's commissioners pledged $4 million in future tax revenues in March 2007 to assist with the purchase and preservation of the 98-acre Angus Tract. Under the terms of Dixon's will, Dixon's widow Edith was permitted to stay on the 23-acre house site ("New" Erdenheim) and retain control of the Natural Lands Trust parcel for up to two years before it fell squarely into the trust's control. At that point, press reports indicated, Peter Williamson, the vice president of conservation services for the trust would open negotiations with Montgomery County to complete the Wissahickon Green Ribbon Trail, something Dixon had resisted to the end. Meanwhile, Dixon's daughter, Ellin Dixon Miller, initially retained control of approximately 100 acres at the Equestrian Tract.[32]

As time ran low on the preservation purchase option, Montgomery County's commissioners also applied for state grant funds from the Department of Conservation and Natural Resources; in July 2007, just a month before the deadline, state officials announced the award of a $1 million keystone grant to assist with the purchase. About this same time, the developers and the estate agreed on a five-month extension of the purchase option to allow more time for funds to be raised.[33] As time ran short, other significant donations came in, including $1.9 million from Whitemarsh Township, leaving the effort just $4 million short of its $14.5 million purchase target.[34] By far the most significant contribution, though, was the last-minute, deus ex machina donation from Merck and Company in

December 2007. As a result of a June 2006 chemical spill that killed many of the fish in the Wissahickon near its Upper Gwynedd plant and temporarily created a fishing ban and water scare for downstream Whitemarsh residents, Merck agreed to a $20 million settlement with state and federal environmental authorities, with $4.5 million of the funds being earmarked explicitly for the purchase of the Angus Tract. This left the fundraising effort just shy of its target.[35]

Finally in December 2008 settlement was completed on the sale of the Angus Tract to the Whitemarsh Foundation.[36] The following year contributions from Whitemarsh Township and the Colonial School District, in addition to a $2 million open space grant from Montgomery County, resulted in the purchase of the Erdenheim Sheep Tract, thus ensuring that at least 300 acres would be permanently preserved.[37] With the addition of Airgas Chairman and CEO Peter McCausland and his wife Bonnie to the mix in 2009, ten separate parties worked out a deal that would protect the remainder of Erdenheim in perpetuity. Calling it the "largest land-conservation deal that has ever taken place in Montgomery County, if not Southeast Pennsylvania," Hugh Moulton, the head of the Whitemarsh Foundation, successfully brought these ten separate parties to the table to conclude the deal. Under the terms of the arrangement, the Whitemarsh Foundation added the Sheep Tract to its existing Angus Tract, the McCauslands purchased 243 acres from the Dixon family (including the sixty-room "New" Erdenheim house on the Main House Tract, a substantial portion of the Trackside Equestrian Tract, a portion of the Sheep Tract, and the Natural Lands Trust Tract), and the Natural Lands Trust secured a permanent conservation easement on all four parcels. Fifteen acres of the Equestrian Tract remained in control of the Dixon family. In subsequent agreements the McCauslands took on responsibility for maintaining the Angus and Sheep Tracts through long-term leases from the Whitemarsh Foundation. Given that the McCauslands were new to farming, they agreed to hire the Dixon family's staff to assist with the transition and help maintain the flocks of sheep and herds of cattle. In addition the various parties agreed to terms that finally allowed completion of the Wissahickon Green Ribbon Trail across portions of Erdenheim's land.[38]

Erdenheim's Future

Today Erdenheim is home to about fifty Angus cattle, dozens of free-range chickens, and a large flock of prize-winning Cheviot sheep. Horses still graze in the paddock and train in the stables, although the one-mile oval lies fallow now, its turf grass grown so high that the rails of the old course and many of the barber poles marking the course's intervals are barely visible at the height of summer. The farm also produces various vegetable crops, including tomatoes, asparagus, and beets, as well as various greenhouse varieties of lettuce, alfalfa, and other greens. Christina Reeves, who became the property manager at Erdenheim in 2009, when the present owners, Peter and Bonnie McCausland, purchased the farm, noted in a 2011 interview that Erdenheim for many years was a "hobby farm," but the new owners intend to make the fruits of their labors more available to the local community. Erdenheim continues to win awards, as well, taking home prizes for its Cheviot sheep at the 2011 Pennsylvania Farm Show. According to Erdenheim's livestock manager in 2011, Phil McMahon, Erdenheim still sells the top 50 percent of its lambs to other breeders.[39]

FACING: Composite map from plates 3, 17, and 20 of *Property Atlas of Montgomery County,* volume D, 1949, from Map Collection, Free Library of Philadelphia, with overlay of current Erdenheim Farm tracts. Overlay boundaries are approximate and for illustrative purposes only.

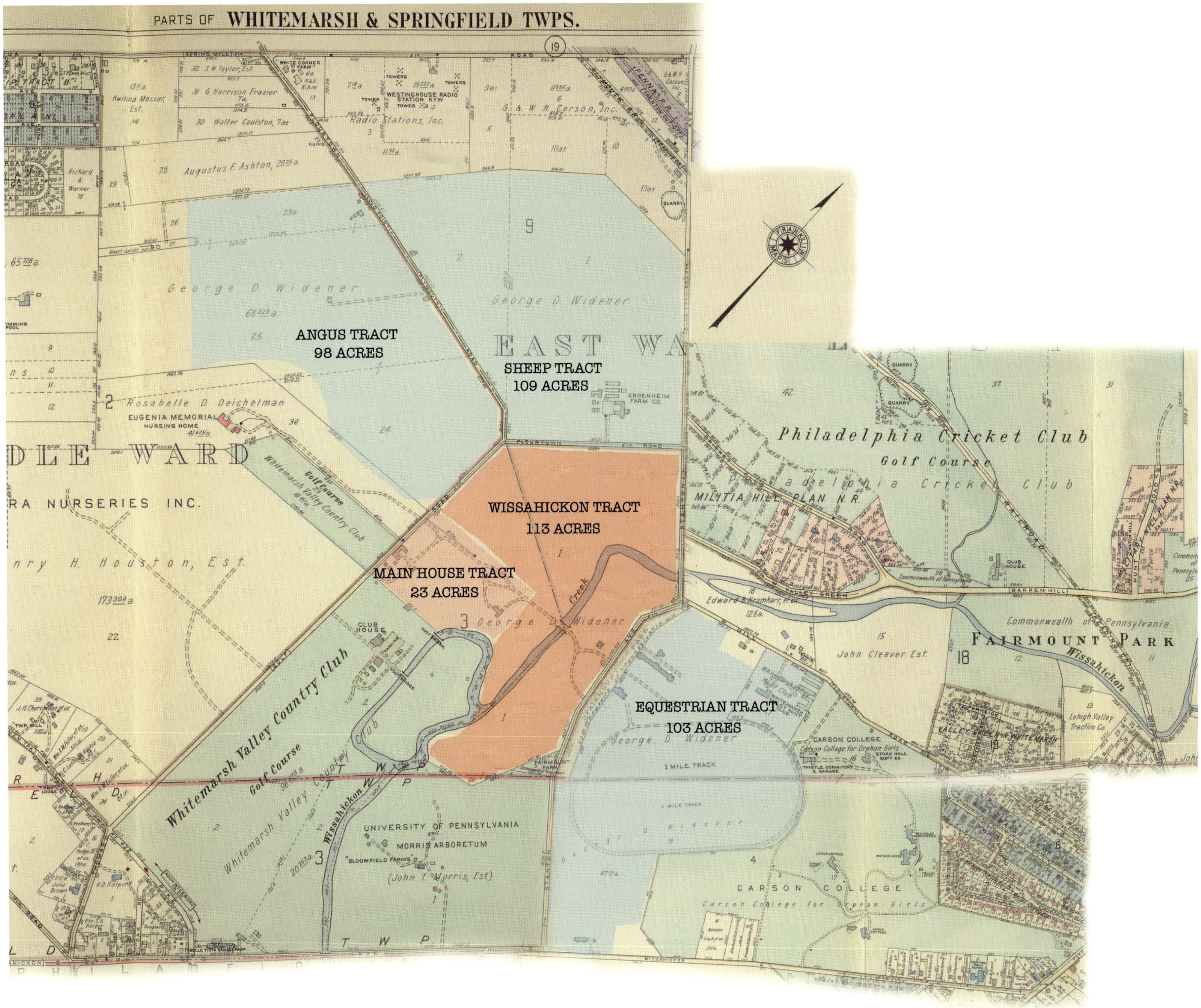

PARTS OF WHITEMARSH & SPRINGFIELD TWPS.
ANGUS TRACT
98 ACRES
SHEEP TRACT
109 ACRES
WISSAHICKON TRACT
113 ACRES
MAIN HOUSE TRACT
23 ACRES
EQUESTRIAN TRACT
103 ACRES
George D. Widener
EAST WA
Philadelphia Cricket Club
Golf Course
MILITIA HILL PLAN N.R.
FAIRMOUNT PARK
Commonwealth of Pennsylvania
Wissahickon
John Cleaver Est.
Whitemarsh Valley Country Club
Golf Course
UNIVERSITY OF PENNSYLVANIA
MORRIS ARBORETUM
CARSON COLLEGE
Carson College for Orphan Girls
1 MILE TRACK
Rosabelle D. Deichelman
EUGENIA MEMORIAL
NURSING HOME
Augustus F. Ashton
WESTINGHOUSE RADIO
STATION KYW
Radio Stations, Inc.
G. & W. H. Corson, Inc.
DLE WARD
RA NURSERIES INC.
nry H. Houston, Est.
ERDENHEIM
FARM CO.
Edward & Krumbhar, et al

But Erdenheim is also a farm in need. Land conservation is one thing, but maintenance and upkeep is quite another. Annual maintenance of the Angus Tract, for example, which is paid for by the Whitemarsh Foundation but performed by the McCauslands and their staff, is estimated at more than $100,000 per year. Stewardship of the Sheep Tract, meanwhile, is paid for and carried out by the McCauslands as part of their lease agreement, but this arrangement could end in 2019, when the lease on that tract expires. As a result the Whitemarsh Foundation continues its efforts today to raise a total endowment of $5 million to maintain the conservation of the Erdenheim Farm lands. These efforts were assisted with a lead gift of $650,000 in July 2009 from Nick and Athena Karabots.[40]

Through these efforts to conserve Erdenheim, its legacy as one of the last remaining open spaces in Montgomery County is assured in perpetuity. But as the farm enters its fifth century as farmland straddling the meandering Wissahickon, there remain questions about its future. Will champions like those in the days of Welch, Kittson, and Widener ever train again on its oval? Will the custodians of Erdenheim be able to restrict development on its fringes, so that its historic vistas are never impeded? Is there a way to recapture—or at least commemorate—what has been lost over the past century at Erdenheim, including the memory of those souls whose final resting places were once marked on the hill overlooking "New" Erdenheim? Is it possible to reconstruct George Hocker's original country estate at "Old" Erdenheim, the sentinel that for so long stood guard over the Wissahickon?

Or, as happens in a world that so frequently forgets the past that made its present, are some things simply lost to the ravages of time?

FACING: The old Lukens house at "New" Erdenheim as it appears today. Photo by Eric Plaag, September 2013.

Notes

1. "Fitz Dixon Jr., Signed Dr J, Dies at 82," *New York Times,* August 5, 2006; James Duffy and Harriet Goodheart, "In Memoriam: F. Eugene 'Fitz' Dixon, Jr.: An Important Friend to Temple University Ambler," press release, August 31, 2006, Temple University, Ambler Campus, http://temple.edu/ambler /news/85-dixon.htm; "Fitz Dixon, 82; Philanthropist Owned 76ers in Late '70s," *Boston Globe,* August 4, 2006; Douglas S. Looney, "That Fitz, He's a Honey," *Sports Illustrated,* November 6, 1978; Platt, "Horace Trumbauer's Mansions," 280–86. Ronaele Manor was demolished in 1973 to make way for over one hundred suburban homes.
2. Looney, "That Fitz."
3. Duffy and Goodheart, "In Memoriam"; "Fitz Dixon, 82; Philanthropist Owned 76ers in Late '70s," *Boston Globe,* August 4, 2006; Andy Wallace, "Fitz Dixon, Former Sixers Owner and Long-Time Civic Leader, Dies," *Philadelphia Inquirer,* August 2, 2006.
4. Douglas S. Looney, "That Fitz, He's a Honey," *Sports Illustrated,* November 6, 1978.
5. Andy Wallace, "Fitz Dixon, Former Sixers Owner and Long-Time Civic Leader, Dies," *Philadelphia Inquirer,* August 2, 2006.
6. "$410,000 for a 2-Year-Old Filly," *Blood-Horse,* March 13, 1972, 882.
7. Terry Conway, "Fitz Dixon: Saying Farewell to a Horse World Giant," *Call to Post,* undated column, http://terryconway.net/cms/index.php?option=com_content&view=article&id=81:fitz-dixon-saying-farewell-to-a-horse-world-giant-&catid=14:racing-columns&Itemid=10.
8. "Fitz Dixon Jr., Signed Dr. J, Dies at 82," *New York Times,* August 5, 2006.
9. Douglas Looney, "That Fitz, He's a Honey," *Sports Illustrated,* November 6, 1978.
10. Duffy and Goodheart, "In Memoriam."
11. "Fitz Dixon, 82; Philanthropist Owned 76ers in Late '70s," *Boston Globe,* August 4, 2006.
12. Looney, "That Fitz."

13. "Fitz Dixon," *New York Times;* Looney, "That Fitz."
14. "An English Countryside Farm Next to Chestnut Hill," *Chestnut Hill Local,* March 31, 2011, http://chestnuthilllocal.com/blog/2011/03/31/an-english-countryside-farm-next-to-chestnut-hill/.
15. "Sheep Show Winners Announced," *Springfield Union,* September 22, 1972, 10.
16. "Fairgoers Undaunted by Weather," *Springfield Union,* September 14, 1972, 3.
17. Douglas S. Looney, "That Fitz, He's a Honey," *Sports Illustrated,* November 6, 1978. The *Philadelphia* magazine article is referenced therein.
18. Jim Foley, "Erdenheim Farm Transitions Gently into the 21st Century," *Challenge,* December 2010.
19. "Fitz Dixon, 82," *Boston Globe.*
20. "Widener University: Campus Life—So Who Were These Guys Anyway?" website, available at http://www.widener.edu,/campuslife/residencelife/oncampus/residencehalls/namesakes.asp.
21. Duffy and Goodheart, "In Memoriam"; "Fitz Dixon, 82," *Boston Globe.* Temple was actually the second taker on the Eleanor Widener Dixon House. Dixon originally offered it to the University of Pennsylvania, who said they might be interested in the offer if he added another $175,000 to cover upkeep for five years. Dixon thereafter remained reluctant to give to the University of Pennsylvania. See Looney, "That Fitz."
22. Duffy and Goodheart, "In Memoriam"; "Fitz Dixon, 82," *Boston Globe;* Andy Wallace, "Fitz Dixon."
23. "Fitz Dixon, 82," *Boston Globe.*
24. Melia Bowie, "Zoning Proposal Discussed, Continued," *Philadelphia Inquirer,* February 16, 2001.
25. Melia Bowie, "Proposed Retirement Complex Rejected," *Philadelphia Inquirer,* March 15, 2001.
26. Chris Gray, "Deal Could Save Much of Montco Farm," *Philadelphia Inquirer,* March 14, 2002.
27. Gray, "Deal Could."
28. Jeff Shields, "Open-Space Proposal Awaits Tax-Plan OK," *Philadelphia Inquirer,* June 24, 2003.
29. Jeff Shields, "Tax Plan to Save Farmland Advances," *Philadelphia Inquirer,* October 20, 2003.
30. Carl Rotenberg, "Kings of the Hill," *Norristown Times Herald,* December 11, 2004.
31. Jeff Shields, "A Swath of Green Needs One Last Link," *Philadelphia Inquirer,* April 20, 2005.
32. Diane Mastrull, "Major Move by Montco to Preserve Erdenheim Farm," *Philadelphia Inquirer,* March 9, 2007; Foley, "Erdenheim Farm"; Jeff Shields, "Philanthropist's Legacy: Green Space, Questions," *Philadelphia Inquirer,* August 31, 2006.
33. Margaret Gibbons, "Montco Applies for $6M Grant," *Norristown Times Herald,* October 2, 2006; Diane Mastrull, "A Victory for Green in Montco," *Philadelphia Inquirer,* July 24, 2007.
34. Diane Mastrull, "Group Nears Goal of Preserving Tract," *Philadelphia Inquirer,* October 5, 2007.
35. Diane Mastrull, "Green by Way of Accident," *Philadelphia Inquirer,* December 15, 2007.
36. Margaret Gibbons, "Merck's $4.5 Million Fine Goes Toward Open Space," *Intelligencer,* December 3, 2008.
37. Margaret Gibbons, "Erdenheim Farm's 450 Acres Preserved as Open Space," *Intelligencer,* May 29, 2009.
38. Whitemarsh Foundation, "History," website, http://whitemarshfoundation.org/farm.htm; Derrick Nunnally, "Big Day for Oasis of Green," *Philadelphia Inquirer,* May 31, 2009; Derrick Nunnally, "Erdenheim Farm Has Changed Hands," *Philadelphia Inquirer,* June 2, 2009.
39. "An English Countryside Farm Next to Chestnut Hill," *Chestnut Hill Local,* March 31, 2011, http://chestnuthilllocal.com/blog/2011/03/31/an-english-countryside-farm-next-to-chestnut-hill/.
40. Whitemarsh Foundation, "Endowment," website, http://whitemarshfoundation.org/saving.htm.

FACING: Dawn and fog rise over the old cemetery hill at New Erdenheim.
Photo by Eric Plaag, September 2013.

FLAME
1980 — 1990
WINSTON
1978 — 1994

BIBLIOGRAPHY

"Abandoned School Sold." *Philadelphia Inquirer,* November 28, 1921, 2.

"Accident on the Marietta." *Sun,* August 29, 1905, 10.

"An Act Not in the Bills." *Philadelphia Inquirer,* August 26, 1880.

"Ad Magraw's Customer: The Son of a Noted Turfman and His Criminal Exploits." *Lancaster Daily Intelligencer,* January 22, 1883.

Advertisement, *Philadelphia Inquirer,* June 16, 1849.

Advertisement, *Philadelphia Inquirer,* May 13, 1906, 13.

Advertisement, *Philadelphia Inquirer,* September 18, 1829.

Aerial Survey of the Philadelphia Region. Regional Planning Foundation of the Philadelphia Tri-State District, c. 1928.

"Agricultural Exhibition." *Farmer's Cabinet and American Herd Book,* August 14, 1847, 25.

"Agricultural Exhibition." *Farmer's Cabinet and American Herd Book,* July 15, 1846, 373.

"A. J. Joyner Dies at Erdenheim." *Blood-Horse,* September 11, 1943, 311–12.

"Album of the Turf." *Blood-Horse,* June 19, 1943, 772.

"All the Turf News: Stryke Takes the Erdenheim Welter Handicap at Morris Park." *Philadelphia Inquirer,* June 8, 1890.

Anderson, Ann. *Snake Oil, Hustlers and Hambones.* McFarland, 2004.

"Ardenheim Farm." *Lewisburg Chronicle.* Reprinted from the *Germantown Telegraph,* August 13, 1852, 1.

"Aristides Welch: The Breeder of Iroquois and Parole." *Sacramento Daily Record Union,* December 13, 1881.

"Atherton Blight." *New York Tribune,* November 5, 1909, 7.

"Atherton Blight Dead." *Philadelphia Inquirer,* November 5, 1909, 9.

Atlas of the County of Montgomery and the State of Pennsylvania. Philadelphia: G. M. Hopkins, 1871.

Atlas of the North Penn Section of Montgomery County, Pennsylvania. Philadelphia: A. H. Mueller, 1916.

Atlas of the Properties in the North-West Suburbs of Philadelphia. Philadelphia: G. Wm. Baist, 1893.

Atlas of the Reading Main Line, Montgomery County, Pennsylvania. Volume 1. Philadelphia: Frank H. M. Klinge, 1927.

Auge, M[oses]. *Lives of the Eminent Dead and Biographical Notices of Prominent Living Citizens of Montgomery County, Pa.* Philadelphia: by the author, 1879.

Baist, G. Wm. *Map of Philadelphia and Environs.* 1889.

———. *Map Showing the Development of the City and Suburbs of Philadelphia.* 1897.

Bateman, Thomas H., compiler. *Widener-Dunton-Elkins-Broomall and Allied Families.* New York: American Historical Company (privately printed), 1953.

Bean, Theodore W., ed. *History of Montgomery County, Pennsylvania.* Philadelphia: Everts and Peck, 1884.

Berg, Denise V. *The Welch Resort: A Hidden Treasure of Boulder County.* Lyons, Colo.: Kumpfenberg, 2006.

"Bert Mulholland on Turf Scene since 1899. *Daily Racing Form,* July 24, 1965.

"A Big Case Settled." *Philadelphia Inquirer,* September 26, 1889, 3.

Bowie, Melia. "Proposed Retirement Complex Rejected." *Philadelphia Inquirer,* March 15, 2001.

———. "Zoning Proposal Discussed, Continued." *Philadelphia Inquirer,* February 16, 2001.

"A Bullet Stilled Her Weary Heart." *Philadelphia Inquirer,* October 16, 1897.

Campbell, John H. *History of the Friendly Sons of St. Patrick and of the Hibernian Society for the Relief of Emigrants from Ireland, March 17, 1771-March 17, 1892.* 1892. Reprint, Kessinger Publishing, 2006.

"Can He Really See That Far Ahead?" N.p., c. 1962, Widener clippings file, Keeneland Library.

"Carson College Will Be Real Fairyland." *Ambler Gazette* [?], May 1917[?].

"Carson Disobeyed Physician's Orders." *Philadelphia Inquirer,* October 17, 1907.

Catalogue of the Fiftieth Annual Exhibition of the Pennsylvania Academy of the Fine Art. Philadelphia: n.p., 1879.

"Chestnut Hill." *Turf, Field, and Farm,* January 6, 1866, 8.

"Chestnut Hill Stud Farm." *New York Herald,* February 22, 1873, 5.

"The Church Home." *Philadelphia Inquirer,* February 5, 1862.

"Circular." *Franklin Gazette,* October 17, 1820.

"City Intelligence." *Philadelphia Inquirer,* January 7, 1864.

"Civil Service Exam Today." *Philadelphia Inquirer,* May 19, 1920.

Coates, Henry T. "Short History of the American Trotting Horse." *Every Horse Owner's Cyclopedia.* Edited by L. C. Bruce. Philadelphia: Porter and Coates, 1880.

Combination Atlas Map of Montgomery County, Pennsylvania. Philadelphia: J. D. Scott, 1877.

"Com. Kittson's Latest." *St. Paul Daily Globe,* January 13, 1882, 2.

"Commencement at Harvard University." *Boston Courier,* July 20, 1854.

"Communication." *Philadelphia Inquirer,* August 27, 1829.

FACING: Horse graves at New Erdenheim. Photo by Eric Plaag, September 2013.

Consolidated Lists of Civil War Draft Registrations, 1863–1865. National Archives and Records Administration. Ancestry.com.
Contosta, David. *Philadelphia's Progressive Orphanage.* University Park: University of Pennsylvania Press, 1997.
Conway, Terry. "Fitz Dixon: Saying Farewell to a Horse World Giant." *Call to Post.* Online blog, http://terryconway.net/cms/index.php?option=com_content&view=article&id=81:fitz-dixon-saying-farewell-to-a-horse-world-giant-&catid=14:racing-columns&Itemid=10.
———. "Leamington: 19th-Century Ruler." *Blood-Horse,* June 19, 2010, 1859.
Cooper, William J., Jr. *Jefferson Davis, American.* New York: Alfred A. Knopf, 2000.
"Corn Planting." *Pennsylvania Farm Journal,* May 1853, 58.
Crist, Lynda Lasswell, ed. *The Papers of Jefferson Davis.* Volume 12. Baton Rouge: Louisiana State University Press, 2008.
Dallin Aerial Survey. *Aerial Survey of Philadelphia, Pa.* 1930.
"Death of Aristides Welch." *Philadelphia Inquirer,* April 10, 1890.
"Death of Aristides Welch." *Turf, Field, and Farm,* April 18, 1890.
"Death of Norman W. Kittson." *New York Times,* May 11, 1888.
"Death of Norman W. Kittson." *Turf, Field, and Farm,* May 18, 1888, 421.
"The Death Roll." *Duluth News-Tribune,* December 4, 1892.
Dewey v. Lambier (1857), 7 Cal. 347.
"Dies in Theatre, Audience Thinks It Fainting Fit." *Philadelphia Inquirer,* October 16, 1907.
"Difenderfer May Be Bowled Out." *Philadelphia Inquirer,* December 28, 1913.
"Drops Dead in Theatre." *New York Times,* October 16, 1907.
Duffin, J. M. *Mapping West Philadelphia: Landowners in October 1777.* University of Pennsylvania University Archives and Records Center website, http://www.archives.upenn.edu/WestPhila1777/about.php.
Duffy, James, and Harriet Goodheart, "In Memoriam: F. Eugene 'Fitz' Dixon, Jr.: An Important Friend to Temple University Ambler." Press release, August 31, 2006, Temple University, Ambler Campus.
"Editors' Book Table." *Godey's Lady's Book,* January 1, 1849, 67.
1880 United States Federal Census. National Archives and Records Administration. Ancestry.com.
1850 United States Federal Census. National Archives and Records Administration. Ancestry.com.
1870 United States Federal Census. National Archives and Records Administration. Ancestry.com.
1860 United States Federal Census. National Archives and Records Administration. Ancestry.com.
"Elect Trainer Bert Mulholland to Hall of Fame at Saratoga." *Daily Racing Form,* May 27, 1967.
E.M. "Local History Sketch." *Ambler Gazette,* October 16, 1902, 2.
———. "Local History Sketch." *Ambler Gazette,* September 1, 1904, 2.
"An English Countryside Farm Next to Chestnut Hill." *Chestnut Hill Local,* March 31, 2011, http://chestnuthilllocal.com/blog/2011/03/31/an-english-countryside-farm-next-to-chestnut-hill/.
"The English Derby." *St. Paul Daily Globe,* June 6, 1881.
"Erdenheim." *St. Paul Globe,* August 24, 1896.
"Erdenheim 'Chase to Fenwick Entry." *Philadelphia Inquirer,* October 10, 1920.
"Erdenheim Events Tomorrow." *Philadelphia Inquirer,* September 7, 1906.
"Erdenheim Farm" advertisement, *Field Illustrated,* December 1920, 1167.
"Erdenheim Farm Familiar Landmark." *Philadelphia Inquirer,* May 9, 1906.
"Erdenheim: Fire at the Famous Stock Farm of Commodore Kittson, near Chestnut Hill." *Philadelphia Inquirer,* October 1, 1884.
"Erdenheim in Flames." *Philadelphia Record* [?], June 8, 1902 [?].
"Erdenheim on Wissahickon." *St. Paul Daily Globe,* December 18, 1882, 2.
"Erdenheim Races Draw Big Crowd, *Philadelphia Inquirer,* October 16, 1921.
"The Erdenheim Sale." *New York Herald,* November 5, 1882.
"The Erdenheim Sale." *New York Herald,* October 17, 1882.
"The Erdenheim Sale." *Turf, Field, and Farm,* November 16, 1888, 460.
"The Erdenheim Sale: Commodore Kittson's Horses at Auction and Fair Prices Obtained." *New York Herald,* October 7, 1886.
"Erdenheim Sold: The Home of Famous Leamington Bought By Robert N. Carson." *Philadelphia Inquirer,* April 30, 1896.
"The Erdenheim Stables." *St. Paul Daily Globe,* December 15, 1884, 2.
"Erdenheim Stock Farm." *Philadelphia Record,* c. 1903.
"Erdenheim Stock Farm." *Sacramento Record-Union,* June 8, 1893, 1.
"Erdenheim Stock Farm—Aristides Welch—Norman W. Kittson—Henry Hocker—Well-Known Property." N.p., c. 1910, Edward Mathews clippings scrapbooks, Historical Society of Montgomery County.
"The Erdenheim Stud." *Field Illustrated,* March 1925.
"Erdenheim Track Races." *Philadelphia Inquirer,* October 20, 1904.
"The Erdenheim Yearlings." *Turf, Field, and Farm,* April 11, 1890, 417.
"Estates of the Dead." *Philadelphia Inquirer,* April 16, 1890.
"Extract of a Letter Dated Harrisburg, March 5." *National Gazette,* March 9, 1824.
"Fairgoers Undaunted by Weather." *Springfield Union,* September 14, 1972, 3.
"Famed Trainer Passes." *Lexington Herald Leader,* July 13, 1968.
Famous Horses of America. Philadelphia: Porter and Coates, 1877.
Farley v. Kittson, 120 U.S. 303 (1887).
"Fight for Income of $4,000,000 Estate." *Philadelphia Inquirer,* March 26, 1909.
Finney, H. S. "Erdenheim." *Blood-Horse,* June 22, 1935, 746.
"1st day of [--] mo, 1727/28." James Steel's Letterbook, 1715–1732, 83, Historical Society of Pennsylvania.
Fish, Larry. "Designed by Trumbauer, Destined for Destruction." *Philadelphia Inquirer,* November 19, 2004.
"Fitz Dixon, 82; Philanthropist Owned 76ers in Late '70s." *Boston Globe,* August 4, 2006.
"Fitz Dixon Jr., Signed Dr J, Dies at 82." *New York Times,* August 5, 2006.

Fitzpatrick, Frank. "Philly's Horse Racing Heritage." *Philadelphia Inquirer Online,* May 19, 2011, http://www.philly.com/philly/blogs/frank_fitzpatrick.
"5000 See Firemen Fight $40,000 Blaze on Widener Estate." *Philadelphia Inquirer,* January 17, 1921. "Flames at Erdenheim." *New York Times,* October 1, 1884.
"Flora Temple's Grave." *Philadelphia Record,* c. 1888.
Foley, Jim. "Erdenheim Farm Transitions Gently into the 21st Century." *Challenge,* December 2010.
Forney, John Wien. *Anecdotes of Public Men.* New York: Harper Brothers, 1873.
"$410,000 for a 2-Year-Old Filly." *Blood-Horse,* March 13, 1972, 882.
Garraty, John A., and Mark C. Carnes, eds. *American National Biography.* Volume 23. New York: Oxford University Press, 1999.
Genealogies of Pennsylvania Families from the Pennsylvania Genealogical Magazine, volume 1: *Arnold-Hertzel.* Ancestry.com.
"General Items." *Blood-Horse,* August 28, 1943, 271.
"The Gentleman Farmer as an Advertiser." *Printer's Ink,* 61:8, November 20, 1907, 17–18.
"George D. Widener, 1889–1971." *Blood-Horse,* December 20, 1971, 4764.
"George Widener, Giant of Thoroughbred Racing, Succumbs." *Mobile Register,* December 9, 1971, 76.
"George Wilkes's Will." *New York Times,* October 10, 1885.
Gibbons, Margaret. "Erdenheim Farm's 450 Acres Preserved as Open Space." *Intelligencer,* May 29, 2009.
———. "Merck's $4.5 Million Fine Goes Toward Open Space." *Intelligencer,* December 3, 2008.
———. "Montco Applies for $6M Grant." *Norristown Times Herald,* October 2, 2006.
Gillingham, Harold Edgar. *The Gillingham Family, Descendants of Yeamans Gillingham.* Philadelphia: Patterson and White, 1901.
"Girard and His Gift Swayed Carson." *Philadelphia Record,* October 27, 1907.
Gray, Chris. "Deal Could Save Much of Montco Farm." *Philadelphia Inquirer,* March 14, 2002.
Hagin, Rab H. "More Francs Than Yen—and Plenty of Dollars." *Blood-Horse,* February 21, 1972, 662, 664.
Hall, Charles. *Geological Map of the Philadelphia Belt of Azoic Rocks.* Harrisburg: Pennsylvania Geologic Survey, c. 1885.
"Harvard University." *Massachusetts Spy,* July 28, 1858.
Hervey, John. *The American Trotter.* New York: Coward-McCann, 1974.
———. *Racing in America, 1922–1936.* New York: Scribner Press, printed privately for the Jockey Club, 1937.
Hildebrandt, Rachel. *The Philadelphia Area Architecture of Horace Trumbauer.* Mount Pleasant, S.C.: Arcadia Publishing, 2009.
Hirsch, Joe. "Tropical Park." *Daily Racing Gazette,* December 23, 1971.
Historic Pennsylvania Church and Town Records. Historical Society of Pennsylvania. Ancestry.com.
Historical Sketches: A Collection of Papers Prepared for the Historical Society of Montgomery County, Pennsylvania. Norristown, Pa.: Herald Printing and Binding Rooms, 1895.
"History of Iroquois: A Good Purchase." *Daily Inter Ocean,* June 2, 1881, 1.
Hocker, Edward W. "'E.M.'—Intensive Historian." In *Historical Sketches: A Collection of Papers Prepared for the Historical Society of Montgomery County, Pennsylvania.* Volume 6, 234–41. Norristown, Pa.: Norristown Press, 1929.
———. "Montgomery County History." *Bulletin of the Historical Society of Montgomery County, Pennsylvania* 12 (Fall 1959–Spring 1961).
Hollingsworth, Kent. "Fillies and Blue Silks." *Blood-Horse,* March 20, 1972, 969.
Holme, Thomas. *A Map of Ye Improved Part of Pennsilvania in America, Divided into Countyes, Tonwships and Lotts . . .* 1687. Reprint from appendix, *Pennsylvania Archives.* Third Series. Map Collection. Pennsylvania Historical and Museum Commission. Pennsylvania State Archives.
Hopkins, G. M. *Atlas of Philadelphia and Environs.* 1877.
"Horrible Suicide." *Patriot,* June 23, 1880.
"Horse Dots." *Louisiana Democrat,* August 7, 1878.
"Horse Racing Giant Widener, Dies at 82." *Lexington Herald,* December 9, 1971.
Hughes, Albion. "Forty Years of Improving the Breed." *Sunday Bulletin Magazine,* April 19, 1964, 6–7.
Hunsicker, Clifton S. *Montgomery County, Pennsylvania: A History.* New York: Lewis Historical Publishing, 1923.
"Hunt Club at Erdenheim Serves Spring Breakfast to Montgomery County Farmers." *Philadelphia Inquirer,* March 27, 1910.
"Is Landis Wrong in His Stand Against Racing in Baseball?" *Augusta Chronicle,* September 10, 1943, 15.
Jackson, Kenneth T., ed. *Dictionary of American Biography.* Supplement 9: 1971–1975. New York: Charles Scribner's Sons, 1994.
"Jamestown." *Turf and Sport Digest,* October 1930.
"Jamestown to Vie in Stakes at Arlington." *Times-Picayune,* March 7, 1931, 12.
Jayne, Horace Howard Furness, ed. *Letters of Horace Furness.* Volume 1. Cambridge, Mass.: Riverside Press, 1922.
"Jerseys Average $1000 at Sale." *Philadelphia Inquirer,* June 6, 1919.
Johnson v. Hocker, 1 U.S. 406 (Dall. 1789).
Jordan, John W., ed. "Jawood Lukens." *Encyclopedia of Pennsylvania Biography.* Volume 1. New York: Lewis Historical Publishing, 1914.
The Judge, "Tom Healey, Trainer." *Turf,* November 15, 1913.
Kelley, Robert F. *Racing in America, 1937–1959.* New York: North River Press, printed privately for the Jockey Club, 1960.
King, Moses. *Philadelphia and Notable Philadelphians.* New York: M. King, 1901.
Kiser, Ellis, Otto Barthel, and B. A. Skeels. *Atlas of Cheltenham, Abington, and Springfield Townships and Vicinity, Montgomery County, Pa.* Philadelphia: A. H. Mueller, 1897.
Kiser, Ellis, and J. M. Lathrop, surveyors. *Atlas of the Properties on the Reading Railway, Embracing*

Cheltenham, Abington, Springfield, and Parts of Moreland and Whitemarsh Townships. Philadelphia: A. H. Mueller, 1909.
"Kittson Was Not Married." *New York Times,* March 4, 1896.
"Lady Lowther." *Washington Times,* May 20, 1909, 7.
Lake, D. J., and S. N. Beers. *Map of the Vicinity of Philadelphia.* Philadelphia: Stone and Pomeroy, 1860.
Land Office Map Collection. Record Group 17, Records of the Land Office. Pennsylvania Historical and Museum Commission. Pennsylvania State Archives.
"Land Sales in Albemarle." *Alexandria Gazette,* September 28, 1857.
Lause, Mark A. *A Secret Society History of the Civil War.* Champaign, Ill.: University of Illinois Press, 2011.
"Letter from Washington." *San Francisco Bulletin,* May 17, 1861.
"List of Articles." Pennsylvania Farm Journal: Devoted to Horticulture, Agriculture, and Rural Economy, November 851, 259.
"List of Judges at the State Fair." *Pennsylvania Farm Journal,* September 1855, 285.
"Local Intelligence." *Philadelphia Inquirer,* February 8, 1862.
"Local Intelligence." *Philadelphia Inquirer,* June 26, 1858.
Looney, Douglas S. "That Fitz, He's a Honey." *Sports Illustrated,* November 6, 1978.
Mackay-Smith, Alexander. *The Race Horses of American: 1832–1872.* Saratoga Springs, N.Y.: National Museum of Racing, 1981.
"Mahlon Sands's Death." *New York Times,* May 9, 1888.
Malone, Dumas, ed. *Dictionary of American Biography.* Volume 10. New York: Charles Scribner's Sons, 1933.
"The Man on the Corner," N.p., September 23, 1909, Springfield Township Historical Society, 2008.96.2.
Map of the Vicinity of Philadelphia from Actual Surveys. Philadelphia: R. P. Smith, 1852.
"Marriage." *Daily National Intelligencer,* March 6, 1856.
Martin, David G. *The Philadelphia Campaign: June 1777–July 1778.* Conshohocken, Pa.: Combined Books, 1993.
Martin, John Hill. *Martin's Bench and Bar of Philadelphia.* R. Welsh, 1883.
"Masters of the Hunt." *Town and Country,* November 9, 1907, 20.
Mastrull, Diane. "Green by Way of Accident." *Philadelphia Inquirer,* December 15, 2007.
———. "Group Nears Goal of Preserving Tract." *Philadelphia Inquirer,* October 5, 2007.
———. "Major Move by Montco to Preserve Erdenheim Farm." *Philadelphia Inquirer,* March 9, 2007.
———. "A Victory for Green in Montco." *Philadelphia Inquirer,* July 24, 2007.
McCrea, James A. "James A. McCrea's Mode of Cultivation." *Pennsylvania Farm Journal: Devoted to Horticulture, Agriculture, and Rural Economy,* February 1853, 347.
McCrea, Jas. A., M.D. "Communications." *Pennsylvania Farm Journal: Devoted to Horticulture, Agriculture, and Rural Economy,* November 1851, 233.
McGuire, Hugh J. "Monday's Widener Dispersal Will Reflect Policy of Man." *Thoroughbred Racing Form,* February 14, 1972.
"Meeting of the State Agricultural Society." *Illustrated New Age,* March 23, 1866.
Merry, Robert W. *A Country of Vast Designs: James K. Polk, the Mexican War, and the Conquest of the American Continent.* New York: Simon and Schuster, 2009.
Miller, Randall M., and William Pencak, eds. *Pennsylvania: A History of the Commonwealth.* University Park: Pennsylvania State University Press, 2002.
"Millionaire Horsemen." *St. Paul Daily Globe,* April 23, 1884, 3.
"Misrepresentation." *United States Telegraph,* October 13, 1827.
M'Liss. "Architect Realizes Dream in Corson [*sic*] College for Girls." *Ambler Gazette* [?], May 1917 [?].
Morris, William E. *Map of Montgomery County, Pennsylvania.* Smith and Wistar, 1849.
"Mr. Kittson's Stable Burned." *New York Times,* October 16, 1885.
"Mr. Kittson to Withdraw." *St. Paul Daily Globe,* August 30, 1886, 9.
"Mr. Longstreth's Death." *North American,* June 4, 1879.
"Mrs. Atherton Blight." *New York Herald-Tribune,* October 13, 1897.
"Mrs. R. N. Carson Buried." Philadelphia Inquirer, July 7, 1912.
"Mrs. Widener Dies in Florida," N.p., March 12, 1968, Widener clippings file, Keeneland Library.
"Mr. Welch's Breeding Farm." *The Spirit of the Times,* November 13, 1869, 202.
"Mr. Welch's Stud Farm." *Spirit of the Times,* June 1, 1872.
"Mr. Widener's Quail." *Sports Illustrated,* October 6, 1958.
Murrell, William. "St. Jacob's Oil." *British Medical Journal,* May 5, 1894, 995.
National Cyclopedia of American Biography. Volume 57. Clifton, N.J.: James T. White, 1977.
National Cyclopedia of American Biography. Volume 4. New York: James T. White, 1917.
"Naval Intelligence." *New York Herald,* August 16, 1856.
Newman, Neil. "Jack Joyner's Life Story." *Blood-Horse,* December 18, 1943, 796–97.
"Newport." *Philadelphia Inquirer,* August 25, 1901.
"New York Letter." *Philadelphia Inquirer,* November 14, 1881.
1900 United States Federal Census. National Archives and Records Administration. Ancestry.com.
"No. 845, Jasper Farmer, Jr., &c." 1683, Philadelphia County Survey Book, letter F, volume 1, 356, Philadelphia City Archives.
"No. 820, Kath. Farmer, 1250 a., Edward Farmer." 1703, Philadelphia County Survey Book, letter F, volume 1, 348, Philadelphia City Archives.
Nolan, J. Bennett, ed. *Southeastern Pennsylvania.* Volume 2. Philadelphia: Lewis Historical Publishing, 1943.
Norris [Edward Hocker]. "A Lafayette Tradition about Erdenheim." *Norristown Times Herald,* undated clipping in Edward Hocker Scrapbooks, Historical Society of Montgomery County.
"Notes about Horses and Trainers." *New York Herald,* December 17, 1891, 3.
Nunnally, Derrick. "Big Day for Oasis of Green." *Philadelphia Inquirer,* May 31, 2009.
———. "Erdenheim Farm Has Changed Hands." *Philadelphia Inquirer,* June 2, 2009.
"An Old Merchant Dying." *North American,* June 3, 1879.

"An Old Stock Farm to Be Revived." *Philadelphia Inquirer,* June 7, 1893.
Olmsted Brothers, "Revised Estimate of Cost of Revising Gardens." February 24, 1925, Olmsted Associates Records, Job Files, 1863–1971, box B417, reel 373, project 7378, Library of Congress.
"Our Washington Correspondence." *Daily Globe* (San Francisco), November 6, 1857.
Palmer, Joe H. "Mulholland and His Horses." *Blood-Horse,* November 20, 1943, 620.
Passport Applications, January 2, 1906–March 31, 1925. National Archives and Records Administration. Ancestry.com.
Passport Applications, 1795–1905. National Archives and Records Administration. Ancestry.com.
Patterson, Theodore Cuyler. "Breeding Thoroughbred Ponies." *Outing, an Illustrated Monthly Magazine of Recreation,* December 1900, 274.
"The Pennsylvania Society for the Prevention of Cruelty to Animals." *Evening Telegraph,* January 18, 1870, 8.
"Pete Wins Match Race with Thos. K." *Philadelphia Inquirer,* November 8, 1908.
Pfitzer, Gregory M. *Popular History and the Literary Marketplace, 1840–1920.* Amherst: University of Massachusetts Press, 2008.
Philadelphia, Pennsylvania, City Directory, 1861. U.S. City Directories, 1821–1989. Ancestry.com
Philadelphia, Pennsylvania, City Directory, 1867. U.S. City Directories, 1821–1989. Ancestry.com.
"A Pioneer of Minnesota," *New York Times,* May 20, 1888.
Platt, Frederick. "Horace Trumbauer's Mansions in Elkins Park." *Bulletin of the Historical Society of Montgomery County* 30:4 (Spring 1997): 252–99.
"Post and Paddock." *Spirit of the Times,* November 10, 1888, 589.
"Probable Murder." *New York Herald,* November 11, 1881.
Property Atlas of Montgomery County, Pa., volume D. Philadelphia: Franklin Survey Company, 1949.
Property Atlas of Montgomery County, Pa., volume D. Philadelphia: Franklin Survey Company, 1938.
"The Purchase of Erdenheim by Commodore Kittson." *St. Paul Daily Globe,* May 29, 1882, 1.
"Racing at Erdenheim." *Philadelphia Inquirer,* August 29, 1909.
"Ra[i]sing Terrapin." *Patriot,* August 30, 1909, 5.
"Reminiscences of the Turf: Imp. Leamington." *Thoroughbred Record,* November 5, 1904, 295.
"Reviewer's Department." *Nineteenth Century: A Quarterly Miscellany,* January 1, 1849, 157.
Rhodemyre, Susan. "Aristides of Pennsylvania." *Thoroughbred Record,* January 23, 1980, 286–87.
Rife, Clarence W. "Norman W. Kittson, a Fur-Trader at Pembina." *Minnesota History* 6 (September 1925): 225–52.
"Robert N. Carson." *Patriot,* March 11, 1911.
Roberts, Ellwood, ed. *Biographical Annals of Montgomery County, Pennsylvania.* New York: T. S. Benham, 1904.
"The Rome of Racers: Erdenheim's Famous Horses Sold at Public Auction." *Philadelphia Inquirer,* November 9, 1888.
Rotenberg, Carl. "Kings of the Hill." *Norristown Times Herald,* December 11, 2004.
"Rotten Eggs and a Conspiracy." *Philadelphia Inquirer,* August 28, 1880.
Ruoff, Henry Wilson, ed. *Biographical and Portrait Cyclopedia of Montgomery County, Pennsylvania.* Part Two. Philadelphia: Biographical Publishing, 1895.
"Sad Events: Two Suicides in Philadelphia in One Day." *Philadelphia Inquirer,* June 23, 1880.
"The Sale of Erdenheim: Aristides Welch's Stock Farm Sold at a Sacrifice." *New York Times,* January 9, 1882.
"Sale of the Erdenheim Stud." *Daily Inter Ocean,* November 9, 1888.
"Sale of the Erdenheim Stud, Property of the Kittson Estate." *Kentucky Livestock Record,* November 17, 1888, 306.
"Scheetz." *Philadelphia Inquirer,* March 23, 1876.
Scull, Nicholas. *A Draught of Springfield Mannor in the County of Philadelphia,* 1732, Land Office Map Collection, Series 17.522, Map #3724, Pennsylvania State Archives.
Seltzer, Ruth. "Widener Entertains at Showplace Farm." *Philadelphia Inquirer,* October 27, 1968.
"Sheep Show Winners Announced." *Springfield Union,* September 22, 1972, 10.
Sherburne, John Henry. *The Life and Character of John Paul Jones, a Captain in the United States Navy.* New York: Adriance and Sherman, 1851. Making of America Collection (online).
Shields, Jeff. "Open-Space Proposal Awaits Tax-Plan OK." *Philadelphia Inquirer,* June 24, 2003.
———. "Philanthropist's Legacy: Green Space, Questions." *Philadelphia Inquirer,* August 31, 2006.
———. "A Swath of Green Needs One Last Link." *Philadelphia Inquirer,* April 20, 2005.
———. "Tax Plan to Save Farmland Advances." *Philadelphia Inquirer,* October 20, 2003.
Shihab, Aziz. *Sirhan.* San Antonio: Naylor, 1969.
Shoemaker, Benjamin D., compiler. *Genealogy of the Shoemaker Family of Cheltenham, Pennsylvania.* Philadelphia: J. B. Lippincott, 1903.
Sloet, Jacob. "Some of the Things Discussed on the Ground with Mr. Gallagher and Mr. Graham." July 22, 1925, Olmsted Associates Records.
"Society Out for Erdenheim Races." *Philadelphia Inquirer,* September 9, 1906, 12.
Sparrow, C. Edward. "Whimsy Gets Erdenheim." *Sun,* October 1, 1915.
"Special Breeding Stock Dispersal for Widener Estate Scheduled Here." *Lexington Herald,* January 7, 1972.
"Sporting Record." *New Hampshire Patriot and State Gazette,* April 14, 1881, 4.
"State Fair—Premiums Awarded." *Pennsylvania Farm Journal,* November 1854, 337–52.
Stillé, Charles Janeway. *Memorial of the Great Central Fair for the U.S. Sanitary Commission.* U.S. Sanitary Commission, 1864.
"Stock Farm Sold at Chestnut Hill." *Philadelphia Inquirer,* April 22, 1916.
"Story of the Globe." *St. Paul Daily Globe,* December 21, 1902, 23.
"Suicide of James E. Kitson [*sic*]." *Washington Times,* November 17, 1900, 8.
Taylor, Agnes Longstreth. *The Longstreth Family Records.* Philadelphia: Press of Ferris and Leach, 1909.
"Thoroughbreds Go Up in Smoke." *Philadelphia Inquirer,* June 8, 1902.
Toll, Jean Barth, and Michael J. Schwager, eds. *Montgomery County: The Second Hundred Years.* Montgomery County Federation of Historical Societies, 1983.

Town and City Clerks of Massachusetts. *Massachusetts, Town Vital Collections, 1620–1988.* Ancestry .com.
"Town and Country Calendar." *Town and Country,* September 10, 1904, 3.
"Trainer Bert Mulholland Suffers Coronary Occlusion." *Daily Racing Form,* August 4, 1967.
"Turf Solon George D. Widener Dies Following Long Illness." *Daily Racing Form,* December 9, 1971.
U.S. IRS Tax Assessment Lists, 1862–1918. National Archives and Records Administration. Ancestry .com.
U.S. Naval Historical Command. *Dictionary of American Naval Fighting Ships,* http://www.history .navy.mil/danfs.
U.S. Passport Applications, 1795–1925. National Archives and Records Administration. Ancestry .com.
Untitled. *Cleveland Plain Dealer,* October 15, 1881.
Untitled. *Daily Inter Ocean,* November 17, 1881.
Untitled. *Montgomery County Law Reporter,* volume 28, 1912, 39–52.
Untitled. *New Hampshire Patriot,* March 4, 1886.
Untitled. *Pennsylvania Gazette,* August 25, 1790.
Untitled. *Pennsylvania Gazette,* June 8, 1791.
Untitled. *Pennsylvania Gazette,* October 27, 1779.
Untitled. *Pennsylvania Gazette,* September 28, 1791.
Untitled. *Pennsylvania Packet,* June 11, 1782.
Untitled. *Reno Evening Gazette,* November 21, 1881
Untitled. *Sartain's Union Magazine of Literature and Art,* April 1, 1849, 288.
Untitled. *Senate Executive Journal,* March 11, 1856, 62.
Untitled. *Spirit of the Times,* April 12, 1890, 517.
Untitled. *Spirit of the Times,* June 11, 1881, 500.
Untitled. *Spirit of the Times,* May 19, 1888, 620.
Untitled. *Spirit of the Times,* November 3, 1888, 560.
Untitled. *Spirit of the Times,* September 7, 1889, 274.
Untitled. *Turf, Field, and Farm,* May 26, 1882, 351.
Untitled. *Washington Reporter,* July 13, 1853.
"Valuable Hunting Horse Stolen from Stock Farm." *Philadelphia Inquirer,* August 7, 1913.
Venable, Elizabeth Marshall. *William Adam Hocker: Justice of the Supreme Court of Florida.* Privately printed, 1941.
V. of Tennessee, "The Horse: Feeding Thoroughbred Yearlings." *Southern Planter,* August 1900, 456.
Vosburgh, W. S. *Racing in America, 1866–1921.* New York: Scribner's Press, privately printed, 1922.
Wallace, Andy. "Fitz Dixon, Former Sixers Owner and Long-Time Civic Leader, Dies." *Philadelphia Inquirer,* August 2, 2006.
Washington, D.C., Marriages, 1826–50. Washington, D.C., District Court. Ancestry.com.
Weber, William H. "Year-Round Flower Production at Erdenheim Farm." *Garden Journal* 11–12 (1961): 16–17.
Weeks, Lyman Horace, ed. *The American Turf: An Historical Account of Racing in the United States.* New York: The Historical Company, 1898.
"What's Going on Here." *Blood-Horse,* February 21, 1972, 661.
"What Was Seen at Erdenheim." *Spirit of the Times,* June 3, 1882, 501.
"Where Robert N. Carson's Millions Will Rear a College for Orphan Girls." *Public Ledger,* October 24, 1907.
Whitemarsh Foundation, "Endowment." http://whitemarshfoundation.org/saving.htm.
Whitemarsh Foundation, "History." http://whitemarshfoundation.org/farm.htm.
"Widener Stallion Seasons Selected." *Blood-Horse,* February 28, 1972, 746.
"Wide Open Race Is Anticipated in 70th Kentucky Derby on May 6." *Trenton Evening Times,* February 16, 1944, 12.
Wilcox, General Cadmus M. *History of the Mexican War.* Edited by Mary Rachel Wilcox. Washington, D.C.: Church News Publishing, 1892. Making of America Collection (online).
Wiley, Samuel, ed. *Biographical and Portrait Cyclopedia of Montgomery County, Pennsylvania.* Philadelphia: Biographic Publishing, 1895.
Will, Johann Martin. *Plan des befestigten Lagers der Amerikaner beÿ Whitemarsh.* Augsburg, Germany: n. p., 1777. Library of Congress, Geography and Map Division.
Williams v. Patterson, 38 Pa. Superior Ct. 322 (1909). *Pennsylvania Superior Court Reports,* volume 38 (1909): 322–26.
"W. P. Thompson, Widely Known Sportsman, Dies." *New York Tribune,* September 21, 1922.
Yeakle, Samuel. *Historical Sketches: A Collection of Papers Prepared for the Historical Society of Montgomery County, Pennsylvania.* Volume 7, 108–15. Norristown: John Hartenstine, 1935.
Yeakle, William R. "The Old Burial-Ground Near Erdenheim." *Bulletin of the Historical Society of Montgomery County* 2:2 (April 1940): 95–104.
"Yearlings of High Degree." *New York Herald,* May 14, 1889.
Young, Billy. "Speaking of Men and Horses." *Cleveland Gazette,* November 18, 1944.